New Employee Safety

Christopher D.B. Burt

New Employee Safety

Risk Factors and Management Strategies

 Springer

Christopher D.B. Burt
Department of Psychology
University of Canterbury
Christchurch, Canterbury
New Zealand

ISBN 978-3-319-18683-2   ISBN 978-3-319-18684-9   (eBook)
DOI 10.1007/978-3-319-18684-9

Library of Congress Control Number: 2015938737

Springer Cham Heidelberg New York Dordrecht London
© Springer International Publishing Switzerland 2015

Printed on acid-free paper

Springer International Publishing AG Switzerland is part of Springer Science+Business Media
(www.springer.com)

Contents

About the Author

Christopher D.B. Burt is an Associate Professor of Industrial and Organizational Psychology, and Director of the Masters in Applied Psychology program, at the University of Canterbury, New Zealand. His current research interests cover the psychological associations between trust and safety, including the relationship between trust and employee's helping behaviors, the development of safety-specific trust in new employees, and the influence of trust on employee safety voicing. He has published a book on managing the public's trust in nonprofit organizations, and over 60 refereed articles in US, British, Australian, and New Zealand journals.

Chapter 1
A Model of New Employee Safety Risks

1.1 Introduction

It is somewhat surprising, given the extent of research reporting new employees have disproportionally more accidents, that there have been few systematic attempts to understand the causal factors associated with new employee accidents. There are research papers across a wide spectrum of subject areas which hint at reasons why new employees are disproportionally represented in accident statistics, but it appears that no systematic attempt has been made to integrate this research into a model. Nor has there been an attempt to formulate a management plan which systematically addresses all of the new employee safety risk factors. It is the aim of this book to address these two issues. New employees have many unique features which individually and collectively increase their chances of being involved in an accident. This book offers a comprehensive understanding of these factors. Each factor is manageable, and a carefully planned approach to new employee safety should drastically reduce new employee accidents.

This chapter provides an overview of the *new employee safety risk model* which I have developed to illustrate the key areas where new employees can be exposed to risk. The model is shown in Fig. 1.1 and is a slight adaption of the model presented in Burt (2014). The model is structured to capture the stages associated with the entry of a new employee into a job/organization. For example, the left-hand side of the model deals with job applicants and recruitment processes, while the far right of the model deals with factors such as familiarization which must occur once a new employee starts a job. Structuring the model to follow the consideration of the safety risks associated with a vacant job, and the path of entry of a new employee into an organization, allows the reader to appreciate how new employee risks can be managed at each stage of the process of going from a job vacancy to a new employee becoming a fully integrated member of the organization. Each chapter in this book is briefly described below and deals with a specific component of the *new employee safety risk model*. Each chapter outlines safety risk factors, the reasons for these risks, and how they might be managed. The final chapter brings all of the

© Springer International Publishing Switzerland 2015
C.D.B. Burt, *New Employee Safety*,
DOI 10.1007/978-3-319-18684-9_1

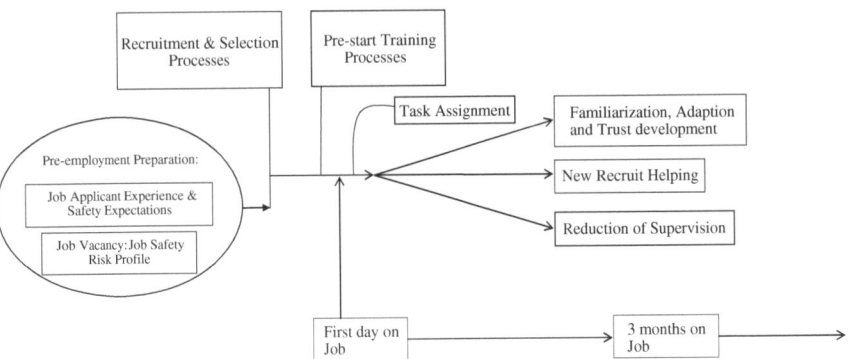

Fig. 1.1 The new employee safety risk model

management suggestions into an overall integrated management plan and describes this plan as a number of steps which can be merged with other human resource management activities which are likely to be occurring in an organization. Before examining the reasons why new employees have accidents, Chap. 2 examines the extent of the new employee safety problem.

1.2 New Employee Accidents

Chapter 2 examines the extensive literature which documents the high accident rate associated with new employees. Research on the relationship between job tenure and accidents, on the relationship between age and accidents, and on the relationship between employee turnover rates and accidents is reviewed. All of the research evidence points to safety issues associated with new employees. Studies from around the world, and conducted in many different industries, have consistently shown higher accident rates associated with new employees. Furthermore, it is very likely that many of the accidents where the injured person is a senior employee (an individual that has worked for the organization for sometime) have in fact been directly caused by a new employee or by factors associated with new employees. Targeting new employee accidents is thus not only necessary to help reduce, and hopefully eliminate, accidents where new employees are injured, but should also help reduce the overall accident rate in an organization.

1.3 Pre-employment Preparation: Job Applicant Experience and Safety Expectations

Chapter 3 discusses different types of job applicant and the assumptions which organizations can, and tend to, make about job applicants. A component which needs consideration before attempting to bring a new employee into an organization is the type of job applicants that are likely to apply for a vacant job. The type of job applicants will be partly determined by the nature of the job, the way in which the job will be advertised, and the selection criteria outlined to applicants. Four different types of job applicant are defined, and the safety implications associated with each type are discussed. Particular attention is paid to the relationship between a job applicant's work experience and their safety expectations. Experience is a complex and perhaps poorly understood concept, and Chap. 3 describes how an understanding of the safety implications of experience is important for the management of new employee safety. New employee safety expectations are built on the foundations of experience, and variation in safety expectations across different types of new employee is discussed.

If an organization considers all new employees are the same or incorrectly assumes that previous work experience provides a new employee with more protection against accidents than it truly does, they are missing an opportunity to reduce new employee accidents. While the organization has the objective of filling a vacant position, and may have a standard approach to the steps they will take, considerable safety advantage can be gained by adjusting processes to address the safety implications associated with each of the four types of job applicant. Chapter 3 offers suggestions on how variation in job applicants' experience and safety expectations can be managed to help address new employee safety risks.

1.4 Pre-employment Preparation: Job Risk Profile

Chapter 4 focuses on the safety risk profile of the job which is being recruited into, and how it is very important that the safety risks associated with the job are well understood, and are not added to by factors associated with new employee arrival. Thus, a further component to consider before beginning to recruit a new employee is the job which is being recruited for and the safety risks associated with the particular job. Clearly, jobs vary in terms of safety risk. Furthermore, while it is theoretically possible to precisely define the safety risks associated with performing a particular task/job, in reality there will be a number of idiosyncratic aspects to a job which preclude generalizing its safety risk from one situation to another. A specific job will have a variable level of safety risk associated with the organization within which the job is performed, the supervision of the job, and the co-workers associated with the job. New employees, to varying degrees, expect the known safety hazards and risks associated with a job, but will have no idea of safety risks which are idiosyncratic to the specific job they are about to enter. For example, it is clear that new employees can behave in ways which increase safety risks. If a new employee is entering a job which is performed in an environment where there are many new employees working, the safety risk level will be substantially different to that which might be normally expected for the job. Chapter 4 discusses hazards and risks which can be added to a job and which can make a job more risky than normal for a new employee. The aim of Chap. 4 is to point out areas which organizations can target to reduce the risks associated with work in the initial period of a new employee's job tenure.

1.5 Recruitment and Selection Processes

One of the boxes in the middle section of Fig. 1.1 is labeled *recruitment and selection processes*. When acquiring a new employee, an organization will undertake a number of recruitment and selection processes which can vary in complexity. While recruitment and selection processes are generally aimed at finding a new employee that can perform the job to a satisfactorily level, they can also consider a job applicant's safety and unfortunately can also have negative safety affects which are generally not well understood by organizations. Chapter 5 discusses recruitment and selection processes and how these can have both a positive and negative impact on new employee safety. This chapter describes how employee's assumptions about recruitment and selection processes are likely to determine how they behave toward a new employee when they begin work. Employees tend to assume that the organization's goal is to recruit and select new employees that can work safely. Furthermore, employees tend to assume this goal will be, or is, achieved. Unfortunately, recruitment and selection processes have serious limitations when comes to ensuring (predicting) new employee safety. Research is discussed in Chap. 5 that suggests that how members of an organization behave toward a new employee is partly determined by the employees' perceptions of the effectiveness of what the organization has done during the recruitment and selection processes. Employees that think recruitment and selection processes have successfully delivered a new employee who will work safely seem to be less inclined to engage in behaviors that will ensure the new employee's, or indeed their own, safety.

The accuracy of assumptions about the effectiveness of recruitment and selection processes will vary with the nature of the processes. Chapter 5 examines a number of recruitment and selection processes, with a particular emphasis on their ability to assess (predict) a new employee's safety behavior. This chapter offers recommendations on the design of recruitment and selection processes and on procedures to ensure employees correctly perceive their organization's ability to predict new employee's safety behavior.

1.6 Socialization and Prestart Training Processes

Often health and safety legislation will require employees to be trained for the work they are undertaking. Associated with this prestart training will be a socialization processes (sometimes referred to as an on-boarding or induction processes) which have general objectives, such as introducing the new employee to the organizations' safety policies and procedures. As with recruitment and selection processes, employees form assumptions about the effectiveness of prestart training and socialization processes. Furthermore, employees adjust their own behavior toward new employees based on these assumptions. Chapter 6 discusses research on socialization and prestart training with an emphasis on the positive safety benefits which these processes can achieve. This chapter also takes a critical look at the limited ability of these processes to achieve positive safety benefits, and on the dangers associated with employees making flawed assumptions about the effectiveness of these processes.

1.7 The Initial Employment Period

Once a new employee has been selected and undergone (more or less) prestart training, they will begin work. This is the beginning of what I will refer to as the *initial period of employment*. The initial period of employment is the period where a new employee is at greatest safety risk, and this can extend for up to one year. While many of the studies examined in Chap. 2 report high accident rates in the first year of working in a job, it is hard to precisely define the period of time during which a new employee is at a significantly increased risk of an accident. Certainly, the initial 3 months of employment in a new job may be a particularly risky time, and it certainly is the time during which adaption, familiarization, and trust development process occur. This risk period applies to a degree whether or not the new employee has had previous work experience. Chapter 7 discusses familiarization, adaption, and trust development processes which occur in the initial period of employment. This chapter also examines processes associated with task assignment and changes in new employee supervision associated with the initial period of employment.

It is a mistake to assume that any new employee can be 100 % prepared for a new job. Every new employee, regardless of their previous employment history, will go through a period of familiarization, a period of adaption, and a period of trust building when they enter a new job. As these processes progress, the new employee's safety risk will reduce, but each of these processes takes time. Furthermore, each process (familiarization, adaption, and trust development) can either be left to run its own course, or can be specifically managed by the organization. In the absence of specific management, the time frame for familiarization, adaption, and trust development will be extended, as will the period when the new employee is at an increased risk of an accident. Chapter 7 describes familiarization, adaption, and trust development and offers suggestions on how these processes can be managed in ways which help ensure the safety of both the new employee and their co-workers.

Several other changes are likely in the initial period of employment. The degree of supervision which a new employee receives will change, and how co-workers behave toward the new employee will change. These changes are partly associated with the development of familiarization, with adaption, and with trust development. Furthermore, these changes have safety implications and, as such, are discussed in Chap. 7.

1.8 Helping Behaviors

Chapter 8 is devoted to a discussion of helping behaviors. While helping others is generally encouraged in society, it can place a new employee at considerable risk. Furthermore, it can also place the person being helped (co-workers) at risk. Unfortunately, new employees may be particularly enthusiastic to show management and co-workers that they are committed to their new job, and to do this, they actively seek to help others whenever they can. While such behavior is undoubtedly motivated by good intentions, it can place the new employee into a situation where they are not equipped to deal with the demands of the situation, or it can place their fellow employees in a situation where activities are happening in the workplace which they were not expecting. Chapter 8 describes a number of different processes associated with helping which can have negative safety consequences. This chapter concludes with suggestions for the management of new employee helping behaviors.

1.9 Measurement

Chapter 9 provides a discussion of scales which can be used to measure a number of the constructs which are discussed in other chapters of the book. This chapter has an emphasis on measurement to provide an organization with a clear understanding of how best to manage new employee safety within their specific context. While a general adoption of the recommendations made in each chapter should go a long way toward ensuring new employee safety, fine-tuning these management strategies with data collected in the specific situation should further enhance the positive benefits. Scales which new employees can complete in order to provide information which will help them understand the safety risks they may face are presented in the first part of this chapter. These are followed by a description of a safety-specific exit survey method which can be used to help generate a safety risk profile for a job. This chapter concludes with a discussion of scales which can be completed by current job incumbents and that provide information which will help job incumbents appreciate the safety risks which new employees may pose.

1.10 An Integrated Management System

Chapter 10 brings all of the management strategies into one place and makes a number of suggestions about how the recommendations made in the other chapters can be integrated with other processes which an organization is likely to use to manage both safety and its human resource. This chapter is presented as a series of steps: An organization will already do more or less at each of these steps, although they may not undertake any activities specifically aimed at improving new employee safety. Chapter 10 attempts to show how new employee safety, and workplace safety in general, can be improved by integrating what are essentially rather simple processes and procedures, into existing human resource activities.

1.11 Conclusions

The aim of this book is to provide a comprehensive discussion of the factors which have the potential to increase safety risks for new employees, and the safety issues which current job incumbents face when new employees arrive. The work is written to provide a research-based understanding of the issues associated with new employee safety risks. As such, students completing courses on occupational health and safety may find this book useful. However, the work will also be useful for managers and practitioners looking for solutions to their new employee accident problems. Recommendations to improve new employee safety are made which can easily be adopted, have relatively little cost, and should easily fit within existing processes.

Reference

Burt, C. D. B. (2014). Managing new recruit safety risks: An integrated model. In R. D. J. M. Steenbergen, P. H. A. J. M. van Gelder, S. Miraglia & A. C. W. M. Vrouwenvelder (Eds.), *Safety, reliability and risk analysis: Beyond the horizon. Proceedings of the 22nd European Safety and Reliability (ESREL, 2013) Conference, Amsterdam* (pp. 1527–1533).

Chapter 2
New Employee Accident Rates

2.1 Introduction

At its broadest level, there are three different bodies of research that have addressed new employees' occupational accident rate. All three literatures clearly show that an employee is more likely to have an accident at work in their initial period of employment in a job. The larger two bodies of literature are those which have examined the relationship between job tenure and accidents, and the relationship between age and accidents. The age literature has tended to focus on young or youth worker, and these workers are often new employees (have relatively short job tenure), but this is not always the case. Generally, the research on the relationship between age and accidents has not attempted to disentangle the relationship between age and job tenure. Nevertheless, and despite some interpretation difficulties, I will examine this literature. Finally, there is a small literature which has looked at the relationship between employee turnover rates and accidents, which is also suggestive of safety issues associated with new employees. Overall, it seems clear that new employees are a safety risk and may even be classified as a safety hazard.

Many safety risks and hazards in workplaces are constant or static. Such risks can be identified, and either removed, guarded against and/or appropriate warnings put in place. That is engineering and ergonomic interventions can help protect employees from known risks and hazards. Furthermore, employees can be trained in ways to cope with or avoid constant or static workplace' safety risks and hazards. In contrast to constant or static safety risks, many organizations have a dynamic workforce which can be relatively constant in size, but continuously changing as people resign and new employees come onboard, or constantly growing in size as operations ramp up, which also sees new employees coming onboard. This flow of new individuals into a workplace can be characterized as a flow of risk and hazard into the workplace in the form of the behaviors and attitudes which the new employee brings to the job.

© Springer International Publishing Switzerland 2015
C.D.B. Burt, *New Employee Safety*,
DOI 10.1007/978-3-319-18684-9_2

A new employee is defined as any individual that has recently started a job. Some research has used the term *newcomer* to describe a new employee (e.g., Molleman and van der Vegt 2007). As will be discussed below, a new employee may also be relatively young (e.g., a youth worker entering their first job), but this is not always the case. The label of *new employee* equally applies to an individual that has previously worked in another job, or in other jobs: They are new to job they enter irrespective of their past employment history. Chapter 3 discusses the relationship between experience and accidents, and describes how even an experienced new employee is still initially a safety risk.

Finally, the research examined below tends to link variables within cases; for example, data on accidents, age, and job tenure are collected from the same employee (sample of employees) and correlated. While this research clearly shows that new employees suffer accidents, it potentially misses the impact of new employees on their co-workers' safety. It is also very likely that some of the accidents suffered by employees that have been working for an organization for some time (what might be termed senior employees) may involve a new employee as part of the causal mechanism. Indeed, responsibility for industrial fatalities/ accidents has been associated with the fellow worker for over 100 years (see Eastman 1910; Swuste et al. 2010). Thus, overall, new employees are both a safety risk to themselves and potentially a safety risk to all employees in an organization.

2.2 Job Tenure and Accidents

A number of different approaches have been taken by researchers examining the relationship between accident statistics and employee job tenure (how long the employee has worked in the job). In some studies, researchers have formed groups of employees based on their job tenure and compared accident rates across the groups. Unfortunately, not all studies that have used this group comparison approach to study the relationship between job tenure and accidents have attempted to control for employee age across the groups. Other studies have used job tenure as a predictor variable in regression analysis or simply correlation analysis in an attempt to find associations between an employee's job tenure and accidents.

A useful example of a study which attempted to identify the unique contribution of a large number of demographic variables (including age) and job-related variables (including job tenure) to workplace accidents was conducted by Leigh (1986). Their analysis used a sample of 4962 draw from the University of Michigan's Panel Study of Income Dynamics for 1978 and 1979, and used logistic regression to analyze relationships. From the perspective of this chapter, a key finding was that the length of time with the job (job tenure) was negatively associated with accidents: Participants with shorter job tenure reported more accidents. Or stated in a different way, new employees had a higher accident rate.

Examples of studies that have looked at accidents within groups defined by employee tenure often focus on specific industries. While such studies may have limited generalizability to other industries, they do tend to focus on the most dangerous occupations/industries and are therefore very valuable in adding to our understanding of occupational accidents. Bennett and Passmore (1984) examined studies conducted on the coal mining industry, noting that at least at the time it was the most dangerous occupation in the US. Bennett and Passmore (1984) reviewed three studies (i.e., Theodore Barry and Associates 1971, 1972; Root and Hoefer 1979) which clearly indicate that job tenure is a significant factor in coal mine accidents. Theodore Barry and Associates (1971) examined a database of 731 fatal underground coal mine accidents and found a strong negative relationship between fatalities and job tenure. Theodore Barry and Associates (1972) examined 688 underground coal mine fatalities and found that in 31 % of the cases, the employee had less than one year of job tenure, and in 7.8 % of the cases, the employee had less than one month of job tenure. Root and Hoefer (1979) examined approximately 270,000 work injuries from ten US states that participated in the Bureau of Labor Statistics Supplementary Data System for 1976 and 1977. Forty percent of the injuries reported had occurred during the first year of employment, and half of these occurred during the employees' first 3 months on the job.

The association between job tenure and accidents in the mining industry is also apparent in more recent studies. Groves et al. (2007) examined Mine Safety and Health Administration (MSHA) and Current Population Survey (CPS) data for equipment-related injuries over the period 1995–2004. Of the 86,398 injuries examined, 28 % occurred to employees in their first year of job tenure, and of the 597 fatalities examined, 31 % occurred to employees in their first year of job tenure. Furthermore, for both injuries and fatalities, the percentages associated with the first year of tenure in a job were by far the greatest identified. At this point, it is important to note that the above studies are not focusing on people in their first job, rather the statistics relate to job tenure, not the participants' overall employment tenure.

Similar patterns of relationship between job tenure and accidents emerge from data relating to other industries. For example, Bentley et al. (2002) reported that 32 % of injuries on logging skid sites occur within the workers' first 6 months of employment. McCall and Horwitz (2005) reported that 51 % of the 1168 trucking accident claims they examined were made by drivers with less than one year of job tenure. Chi et al. (2005) found that 80.5 % of the 621 fatal occupational falls in the Taiwanese construction industry which they analyzed had occurred in the individual's first year on the job. Jeong (1998) examined national statistics on industrial accidents in the construction sector in South Korea in the years 1991–1994 and found that 95.6 % of the 120,417 non-fatal injuries and 92.5 % of the 2,803 deaths examined had occurred in the employees' first year on the job. Also see Bell and Grushecky (2006), Cellier et al. (1995), Haller et al. (2009), and Kincaid (1996) for other data showing that new employees have higher accident rates when compared to more senior employees.

Much of the research on the relationship between job tenure and accidents has reported or is suggestive of a negative linear trend. That is as an employee's job tenure increases, their likelihood of an accident decreases. But there are exceptions.

Keyserling (1983) used a quasi-experimental design to explore the relationship between job tenure and accidents, by grouping employees into several categories. They found that individuals in a probationary employee group (those with less than 3 months of job tenure) and individuals in an experienced worker group (those with a minimum of one year of job tenure) had fewer accidents than employees in another group with 3 and 12 months of job tenure. They suggested that the relationship between job tenure and accidents is better characterized as an inverted U-shaped relationship, rather than a negative linear relationship. This is an important finding. In particular, it raises the very important question of what happened in the first 3 months of employment that 'protected' new employees in their sample from accidents? Perhaps, in this organization, new employee safety was actively managed in their initial period of employment.

Figure 2.1 shows three hypothetical relationships between job tenure and accidents. Clearly, there is evidence to support the possibility of both a negative linear relationship and an inverted U-shaped relationship between job tenure and accidents. Furthermore, it is likely that organizations may experience either, or both of these relationships. The ideal situation is shown by the dotted line in Fig. 2.1, where

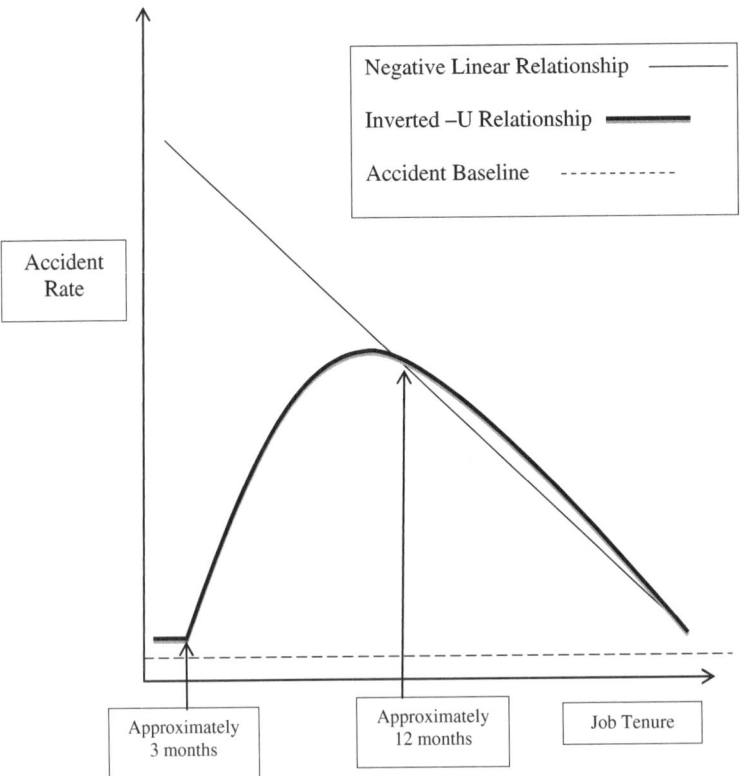

Fig. 2.1 Hypothetical relationships between job tenure and accident rates

the accident rate of new employees is no more than the base level of accidents (hopefully zero) which the organization or specific occupation has historically recorded. The overall objective of this book is to help organizations achieve this zero accident rate outcome for new employees.

While the weight of research evidence seems to overwhelmingly show a relationship between job tenure and accidents, and this book will offer a number of reasons for this finding, it is worth noting that some of the difference in accident rates between new and more senior employees could be due to a sampling bias problem. There is a possibility that the occurrence of accidents has a selection effect on employees. That is employees that have accidents early in their employment may remove themselves from the employment because they are injured or have pressure put on them by family members to get a safer job, etc. This selection process may result in more senior workers (those with longer job tenure) being a selected group of employees (perhaps a group that are very safety conscious). Such a selection bias could exaggerate the difference in accident rates in studies which have used cross-sectional designs to compare accident rates between new employees and more senior employees. While this selection bias probably does occur, it really only explains why senior employees potentially have fewer accidents, not why new employees have many accidents.

2.3 Employee Age and Accidents

One of the most comprehensive reviews of young workers' (under 25 years of age) occupational accidents was conducted by Salminen (2004) (also see Castillo 1999; Rhodes 1983; Laflamme and Menckel 1996; Salminen 1996 for earlier reviews). Salminen's review was interested in two key questions: Do young workers have higher rates of occupational injuries? And do young workers have more fatal accidents? Salminen identified 63 studies which had addressed the first question and 45 studies relevant to the second question. The studies as a whole were conducted in many different countries (i.e., America, UK, Japan, Holland, Sweden, Israel, New Zealand, Australia, Canada, Finland, Germany, Denmark, France, Jordon, Norway, Brazil, China, Taiwan, and Iceland). Fifty-six percent of the studies on non-fatal injuries found that young workers had a higher injury rate than older workers, whereas 64 % of the studies on fatal injuries showed that the rate was lower in young workers. Collectively, the data examined in Salminen's review support the notion that young workers are more likely to be injured at work, although thankfully this injury may not be fatal. As might be expected, there was considerable variation in findings across the studies and also across the industries where the studies are sampled. While the interested reader can consult the review for the specific details—the message is clear, organizations need to give consideration to the very strong possibility that young workers will be involved in an accident.

Salminen's (2004) review does not attempt to provide many detailed explanations for their findings. That is they do not attempt to explain why young workers suffer more non-fatal accidents (apart from suggesting that young workers may lack experience or are less likely to be killed by an impact which would probably kill an older worker—the latter not being a very comforting explanation). The review also does not address whether the young workers (classified as aged less than 25 years in their review) were in fact new employees. While it is probably reasonable to assume that many young workers in the studies reviewed were in fact new employees, some may have already been working for several years (possibly up to 10 years depending on the minimum school leaving age in the various countries represented by the reviewed studies).

The difficulty associated with disentangling the age/job tenure relationship in many studies is due to the large age ranges that are often used to form the youth worker or young worker group. For example, Laflamme (1996) who examined aged-related accident risks in the Swedish Automobile industry using a well-designed retrospective longitudinal study identified higher accident rates among young worker (aged 16–24 years), but failed to account for the potential of an 8-year within-group job tenure difference. Of course, this is not always the case. For example, Scott et al. (2004) reported that Australian youth workers in the 15–17 years of age group were twice as likely to experience a work-related injury as other workers, and it might be reasonable to assume that job tenure for this group was relatively short, given that the youngest age at which an individual can leave school in Australia ranges from 15 to 17 depending on state.

Other results which illustrate the likelihood that many accidents associated with age are occurring because the employee can be classified as a new employee are provided by the studies of Lin et al. (2008) and Van Zelst (1954). Lin et al. (2008) found that males aged 24 years or less had the highest rate of fatal occupational injuries in an analysis of 1890s accident reports filed between 1996 and 1999 in Taiwan. However, the truly revealing statistic in Lin et al.'s (2008) study was the finding that when length of work experience (job tenure) was known, which was the case in a total of 977 of the 1890s accidents examined, 61.5 % of the fatal accidents had occurred during the first year of employment.

Van Zelst (1954) examined a group of employees ($N = 297$) that had a mean age of 39.2 years but who were <u>inexperienced</u> in the type of work being performed (they were new employees) and found a higher-than-normal accident expectancy rate in their initial period of employment. The normal accident expectancy rate was defined as the organization's baseline level of accidents. Thus, being older did not seem to remove all safety risks associated with being a new employee. However, the study also found that the group's accident rate did reach the normal accident expectancy level at approximately 2 months of job tenure. Chapter 3 discusses the generalizability of workplace experience from one job to another, which may partly explain Van Zelst (1954) finding.

In summary, there are characteristics associated with age which can increase the possibility of an accident. For example, the anthropometric characteristics of youth and adults are different, and machinery may well have been designed to accommodate the anthropometric characteristics of an adult working population. Furthermore, there are also a number of psychological attributes associated with youth which can increase the possibility of an accident: poor judgment, sensation-seeking, poor risk assessment, vulnerability to peer pressure, incomplete self-image, pressure to excel, proving one's independence and maturity, and a need to rebel tend to be characteristics associated with youth. However, it is also the case that many youth workers are also likely to be new employees. Throughout this book, it will be argued that a lot of the safety risks which youth (young) workers experience are due to their new employee status, rather than specifically because of their age or factors associated with their age.

2.4 Employee Turnover Rates and Safety

Given that there is clear evidence that employee job tenure and accident rates are associated, it might be expected that there would be a body of research evidence showing a relationship between employee turnover rates and accident/incidence occurrence. That is organizations that have high voluntary turnover, where the employee leaves and the organization replaces them, are likely to have high accident rates associated with the volume of new employees entering the workforce. There is of course a reasonable volume of research on the factors associated with employee turnover. However, the safety factors associated with employee turnover appear to be rarely mentioned.

The research evidence on the relationship between employee turnover and organizational performance is somewhat mixed and has tended to focus more on the good effects of employee turnover on organizational performance. For example, it is possible that a low-to-moderate degree of turnover may be good for an organization, in that the low-to-moderate level of turnover may be sufficient to remove poor performers (Abelson and Baysinger 1984), introduce new knowledge and skills (Alexander et al. 1994), and reduce employee homogeneity and increase diversity (Schneider et al. 1995). However, throughout this book, I will argue that any level of employee turnover can potentially be negative for safety performance.

While the vast majority of research on employee turnover has not examined its impact on safety, Shaw et al. (2005) is a notable exception. Furthermore, Shaw et al.'s (2005) study appears to have prompted safety to be included as a performance dimension in recent models of the relationship between employee turnover and organizational performance (e.g., see Fig. 2.1 in Shaw 2011). Shaw et al.'s (2005) study suggested four alternative relationships between employee voluntary turnover and organizational performance (with safety included as one dimension of performance): linear negative relationship, inverted U-shaped relationship, attenuated negative relationship, and the HRM-moderated relationship. For the linear negative relationship, Shaw et al. (2005) used Staw's (1980) suggestion that high turnover would deplete the resources that are available to do what might be termed *optional activities* such as maintenance, and the lack of these optional activities could have a negative impact on safety. In the case of an inverted U-shaped relationship, the general argument is that a low level of turnover is good for organizational performance (as noted above). While Shaw et al. did not speculate how this low level might be good for safety, it is possible that a new employee (despite the many risks they bring, as will be discussed in other parts of this book) might also bring with them new safety ideas. The 'attenuated negative effects of turnover on performance relationship' basically refers to the prediction that as the turnover rate increases, the human capital loss associated with each individual that leaves is lowered. Put simply, the individual leaving probably has not been there that long anyway (given the high turnover rate) and as such has not acquired that much organization-specific human capital. Thus, the negative impact of their voluntary turnover on organizational performance is reduced (attenuated) by the high overall rate of turnover (their short job tenure). In terms of safety, if the organization has a continuously high turnover rate, its base level of accidents may be rather high—lots of new employees and lots of accidents. Finally, in the case of a HRM-moderated relationship between turnover and performance, it is argued that the consequences of turnover on performance vary as a function of HRM practices. In relation to safety, the material in this book relating to practices and processes to reduce new employee safety risks should result in an HRM-moderated relationship between employee turnover and safety performance. Overall, no matter how one looks at employee turnover, there is clearly potential for turnover (and the associated arrival of new employees) to negatively impact workplace safety.

2.5 Can the Problem of New Employee Safety Risks Get Worse?

There are a number of reasons why it might be expected that the number of new employees in organizations might steadily increase. While the safety issues associated with new employees, without any major changes to current practices, are likely to remain stable (the same), accident statistics associated with new employees are likely to show increases simply because there are likely to be more new employees entering organizations. In the following four sections, reasons why the number of new employees is likely to increase are examined: predicted global employee turnover rate increases; retirement of the baby boomer generation; the nature of the contemporary workforce; and recovery from the global recession.

2.5.1 Predicted Global Employee Turnover Rate Increases

Evidence is mounting which suggests that employee turnover rates are steadily increasing. A paper published by the Hay Group (2012) based on research in association with the Centre for Economic and Business Research (CEBR) claimed that globally, the number of workers leaving their jobs is expected to have reached 161.7 million by 2014. Furthermore, the paper claims that as the global economic recovery takes hold, dissatisfied workers will take the opportunity to change jobs. The Asia-Pacific region was specifically noted as likely to see a turnover rate increase from around 21.5 % in 2012 to 25.6 % by 2018. The paper also listed expected turnover rates for 2013 in India of 26.9 %, Russia 26.8 %, Indonesia 25.8 %, Brazil 24.4 %, US 21.8 %, China 21.3 %, and UK 14.6 %. Clearly, these figures not only point to millions of employees leaving their job, but also imply that millions of individuals are going to become new employees.

While it is clear that the predicted employee turnover rates mentioned above are across the entire employment sector, and some of the industries represented by these statistics may not have much in the way of work-related safety issues (e.g., an employee working in a service job is generally exposed to a less dangerous work context, compared to an employee working in mining), a number of industries that operate in high-risk work situations have reported difficulties retaining newly hired employees. For example, Delgoulet et al. (2012) reported this was the case in the construction sector in France.

Many reasons have been offered for why employees do not stay with their employer, with perhaps the single most dominant cause being dissatisfaction with some aspect of the employment situation (e.g., see models by Griffeth et al. 2000; Hom and Kinicki 2001). The centrality of satisfaction is also noted in several reviews of turnover-associated variables (e.g., Burt 2014; Holtom et al. 2008; Hom et al. 2012; Park and Shaw 2013; Shaw 2011; Shaw et al. 2005). One aspect of the employment context which an employee may be dissatisfied with is safety. That is one factor which is known to increase turnover intentions, and ultimately employee turnover, is employee's safety risk perceptions (Cree and Kelloway 1997). Thus, while employee turnover may 'cause' safety problems associated with new employees, it is also possible that safety issues can also cause employee turnover. Interestingly, or perhaps surprisingly, there seems to have been very little research attention given to the possibility that dissatisfaction with safety may be a motivating factor for voluntary turnover.

It is possible that employee turnover and the increased safety risks associated with new employees, starts a cycle or even what might be termed a new employee associated risk avalanche. If we assume that a new employee is a safety risk (and as outlined above, there is very good research evidence to support this assumption), their addition to the workplace may increase the risk perceptions of other employees. That is if an employee sees new employees being injured, killed, or creating hazardous work circumstances for other employees, their perceptions of safety risks associated with the work may increase, and their satisfaction with workplace safety may decline. This may lead to the employee resigning from their job, creating the need to employ a new employee, and that new employee further increases the risk perceptions of other employees, leading to further resignations and so on.

It really is unsurprising that employees who perceive their work or their workplace is unsafe (risky) begin to consider whether it is a good idea to remain in the job. Hirschman (1970) proposed the 'exit–voice' model of the response to high safety risk. Put simply, employees faced with unacceptable safety risk 'talk with their feet'—they leave. The exiting employee creates a vacancy, and in filling this vacancy, the organization may be adding more risk to the workplace in the form of another new employee. This could be characterized as a cycle where risk prompts resignations, which prompts the recruitment of new employees, which increases perceived risk, and so on. Eventually, if the cycle is not broken, the organization may face a risk avalanche where the workplace is predominantly staffed with new employees, and the accident rate is increasing.

2.5.2 Retirement of the Baby Boomer Generation

The baby boomer generation, undoubtedly well over a 100 million people globally (with roughly 78 million in the US alone, Callanan and Greenhaus 2008), who were born in the 20 years post–World War II, are now entering retirement age (65). While there are many efforts to keep this valuable human resource in the workplace (e.g., Callanan and Greenhaus 2008; Dohm 2000), the baby boomers will inevitably retire at some stage. A significant volume of literature has addressed the labor shortages which will be associated with the retirement of the baby boom generation (e.g., Cappelli 2005; Dohm 2000; Lewis and Cho 2011). While some authors have noted how the retirement of baby boomers might be good for organizations, through the introduction of new ideas and motivated staff, there clearly are some safety disadvantages associated with an influx of new employees.

One major disadvantage that is very clear (although there appears to be little if any research literature on it) is that where possible, and for some jobs, in some places, there are likely to be labor shortages, and the knowledgeable, skilled, and experienced baby boomers are going to be replaced with substantially less knowledgeable, skilled, and experienced individuals. Put simply, retiring baby boomers are going to be replaced with new employees. Clearly, from the perspective of this book, the mass retirement of the baby boomers over the next couple of decades is going to see a roughly equally inflow of new employees into workplaces. Given the statistics reported above, it is clear that without well-considered interventions, the retirement of the baby boomers is going to be associated with an increase in workplace accidents, death, and injuries.

2.5.3 Nature of the Contemporary Workforce

The twenty-first century has seen a change in the way organizations form their workforce. What might be termed the contemporary workforce is somewhat different to traditional work/employment relationships, with more emphasis being placed on short-term, temporary, or fixed-term contracts. Associated with this is an increase in the frequency of changes between jobs and workplaces (Papadopoulos et al. 2010). Indeed, it has been reported that individuals change jobs 10.2 times on average in each 20-year period (Bureau of Labor Statistics 2005). Clarke (2003) discussed in some detail the impact of the contemporary workforce on the development of organizational safety culture, noting how safety culture requires (among other things) an opportunity to develop relationships based on trust (see Chap. 7 for a further discussion of trust development). Clarke argued that a workforce that is characterized by short tenure (a contemporary workforce) is going to face difficulties in achieving the degree of integration and interaction required for safety culture development. Koukoulaki (2010) also gives an excellent overview of a number of effects on safety that may result from the changing work environment.

All of the characteristics of the contemporary workforce point to an increase in new employees in organizations. If an organization uses short-term temporary, or fixed-term, contracts, they are creating a continuous or semi-continuous flow of new employees into the organization. The same can be said for the use of project-based employment, where individuals are employed for a specific project, or for the use of subcontracting, or for the use of temporary agency staff. While such staffing arrangements may show cost saving on one line of a balance sheet, they may also increase costs associated with accidents, which in the long term may reduce the apparent savings to zero, or even make it more costly to adopt a contemporary workforce arrangement.

2.5.4 Recovery from the Global Recession

Clearly, there are parts of the world where the global economic recession has not fully lifted. However, other countries are showing positive economic growth, decreases in unemployment rates, and increases in the creation of new jobs. It is likely that economic growth will continue to increase and spread to other parts of the globe. Associated with this recovery will be vast numbers of individuals entering employment, becoming new employees. Thus, while productivity gains are likely for many sectors, increases in accidents due to increases in the number of new employees are also likely.

2.6   Conclusions

Despite variation in study methodology, and causation around potentially confounding issues, the overwhelming weight of evidence clearly indicates that the likelihood of an accident is greatest in the initial period of employment, when the individual is a new employee. There are also a number of reasons why the proportion of new employees in organizations is likely to increase. Unless a systematic approach is taken toward managing new employee safety risks, an increase in accidents will parallel the arrival of new employees. The remaining chapters of this book examine a number of factors which may contribute to new employees' high accident rate, and offer suggestions for the management of each factor.

References

Abelson, M. A., & Baysinger, B. D. (1984). Optimal and dysfunctional turnover: Towards an organizational level model. *The Academy of Management Review, 9*, 331–341.

Alexander, J. A., Bloom, J. R., & Nuchols, B. A. (1994). Nursing turnover and hospital efficiency: An organizational-level analysis. *Industrial Relations, 33*, 505–520.

Bell, J. L., & Grushecky, S. T. (2006). Evaluating the effectiveness of a logger safety training program. *Journal of Safety Research, 37*, 53–61.

Bennett, J. D., & Passmore, D. L. (1984). Correlates of coal mine accidents and injuries: A literature review. *Accident Analysis and Prevention, 16*(1), 37–45.

Bentley, T. A., Parker, R. J., Ashby, L., Moore, D. J., & Tappin, D. C. (2002). The role of the New Zealand forestry industry surveillance system in a strategic ergonomic, safety and health research programme. *Applied Ergonomics, 33*, 395–403.

Bureau of Labor Statistics. (2005). Retrieved 17th February 2015 from http://www.bls.gov/nls/nlsfaqs.htm#anch.

Burt, C. D. B. (2014). Job Satisfaction. In P. Flood & Y. Freeney (Eds.), *the organizational behaviour vol of the wiley encyclopaedia of management* (3rd ed.). UK: Wiley.

Callanan, G. A., & Greenhaus, J. H. (2008). The baby boom generation and career management: A call to action. *Advances in Developing Human Resources, 10*(1), 70–85.

Cappelli, (2005). Will there really be a labor shortage? *Human Resource Management, 44*, 143–149.

Castillo, D. N. (1999). Occupational safety and health in young people. In J. Barling & E. K. Kelloway (Eds.), *Young workers: Varieties of experience* (pp. 159–200). Washington, DC, USA: American Psychological Association.

Cellier, J. M., Eyrolle, H., & Bertrand, A. (1995). Effects of age and level of work experience on occurrence of accidents. *Perceptual and Motor Skills, 80*(3), 931–940.

Chi, C., Chang, T., & Ting, H. (2005). Accident patterns and prevention measures for fatal occupational falls in the construction industry. *Applied Ergonomics, 36*, 391–400.

Clarke, S. (2003). The contemporary workforce: Implications for organizational safety culture. *Personnel Review, 32*(1), 40–57.

Cree, T., & Kelloway, E. K. (1997). Responses to occupational hazards: Exit and participation. *Journal of Occupational Health Psychology, 2*(4), 304–311.

Delgoulet, C., Gaudart, C., & Chassaing, K. (2012). Entering the workforce and on-the-job skills acquisition in the construction sector. *Work, 41*, 155–164.

Dohm, A. (2000). Gauging the labor effects of retiring baby boomers. *Monthly Labor Review, 123*(7), 17–25.

Eastman, C. (1910). *Work-accidents and the law. The Pittsburg survey*. New York: Charities Publications Committee.

Griffeth, R. W., Hom, P. W., & Gaertner, S. (2000). A meta-analysis of antecedents and correlates of employee turnover: Update, moderator tests and research implications for the next millennium. *Journal of Management, 26*(3), 463–488.

Groves, W. A., Kecojevic, V. J., & Komljenovic, D. (2007). Analysis of fatalities and injuries involving mining equipment. *Journal of Safety Research, 38*, 461–470.

Haller, G., Myles, P., Taffé, P., Perneger, T. V., & Wu, C. L. (2009). Rate of undesirable events at beginning of academic year: Retrospective cohort study. *BMJ, 2009*(339), b3974. doi:10.1136/bmj.b3974.

Hay Group. (2012). http://www.haygroup.com/downloads/in/Retention%20study%20India%20press%20release%20Final.pdf. Accessed 10 April 2014.

Hirschman, A. O. (1970). *Exit, voice, and loyalty*. Cambridge, MA: Harvard University Press.

Holtom, B. C., Mitchell, T. R., Lee, T. W., & Eberly, M. B. (2008). Turnover and retention research: A glance at the past, a closer review of the present, and a venture into the future. *The Academy of Management Annals, 2*(1), 231–274.

Hom, P. W., & Kinicki, A. J. (2001). Toward a greater understanding of how dissatisfaction drives employee turnover. *The Academy of Management Journal, 44*(5), 975–987.

Hom, P. W., Mitchell, T. R., Lee, T. W., & Griffeth, R. W. (2012). Reviewing employee turnover: Focusing on proximal withdrawal sates and expanded criterion. *Psychological Bulletin, 138*(5), 831–858.

Jeong, B. Y. (1998). Occupational deaths and injuries in the construction industry. *Applied Ergonomics, 29*(5), 355–360.

Keyserling, W. M. (1983). Occupational injuries and work experience. *Journal of Safety Research, 14*, 37–42.

Kincaid, W. H. (1996). Safety in the high-turnover environment. *Occupational Health and Safety, 65*, 22–25.

Koukoulaki, T. (2010). New trends in work environment—New effects on safety. *Safety Science, 48*, 936–942.

Laflamme, L. (1996). Age-related accident risks among assembly workers: A longitudinal study of male workers employed in the Swedish automobile industry. *Journal of Safety Research, 27*(4), 259–268.

Laflamme, L., & Menckel, E. (1996). Age and occupational accidents in the light of fluctuations on the labor market: The case of Swedish non-ferrous ore miners. *Work, 6*, 97–105.

Leigh, J. P. (1986). Individual and job characteristics as predictors of industrial accidents. *Accident Analysis and Prevention, 18*(3), 209–216.

Lewis, G. B., & Cho, Y. J. (2011). The aging of the state government workforce: Trends and implications. *The American Review of Public Administration, 41*(1), 48–60.

Lin, Y., Chen, C., & Luo, J. (2008). Gender and age distribution of occupational fatalities in Taiwan. *Accident Analysis and Prevention, 40*, 1604–1610.

McCall, B. P., & Horwitz, I. B. (2005). Occupational vehicular accidents claims: A workers compensation analysis of Oregon truck drivers 1990–1997. *Accident Analysis & Prevention, 37*(4), 767–774.

Molleman, E., & van der Vegt, G. S. (2007). The performance evaluation of novices: The importance of competence in specific work activity clusters. *Journal of Occupational and Organizational Psychology, 80*, 459–478.

Papadopoulos, G., Georgiadou, P., Padazoglou, C., & Michaliou, K. (2010). Occupational and public health and safety in a changing work environment: An integrated approach for risk assessment and prevention. *Safety Science, 48*, 943–949.

Park, T., & Shaw, J. D. (2013). Turnover rates and organizational performance: A meta-analysis. *Journal of Applied Psychology, 98*(2), 268–309.

Rhodes, S. R. (1983). Age-related differences in work attitudes and behaviors: A review and conceptual analysis. *Psychological Bulletin, 93*(2), 328–367.

Root, N., & Hoefer, M. (1979). The first work-injury data available from new BLS study. *Monthly Labor Review, 102*(1), 76–80.

Salminen, S. (1996). Work-related accidents among young workers in Finland. *International Journal of Occupational Safety and Ergonomics, 2*(4), 305–314.

Salminen, S. (2004). Have young workers more injuries than older ones? An international literature review. *Journal of Safety Research, 35*, 513–521.

Schneider, B., Goldstein, H. W., & Smith, D. B. (1995). The ASA framework: An update. *Personnel Psychology, 48*, 747–773.

Scott, D., Hockey, R., Barker, R., Sprinks, D., & Pitt, R. (2004). Half the age-twice the risk: Occupational injury in school age children. *Injury Bulletin, 84*, 1–4.

Shaw, J. D. (2011). Turnover rates and organizational performance: Review, critique and research agenda. *Organizational Psychology Review, 1*, 187–213.

Shaw, J. D., Gupta, N., & Delery, J. E. (2005). Alternative conceptualizations of the relationship between voluntary turnover and organizational performance. *Academy of Management Journal, 48*(1), 50–68.

Staw, B. M. (1980). The consequences of turnover. *Journal of Occupational Behavior, 1*, 253–273.

Swuste, P., van Gulijk, C., & Zwaard, W. (2010). Safety metaphors and theories, a review of the occupational safety literature of US, UK and The Netherlands, till the first part of the 20th century. *Safety Science, 48*, 1000–1018.

Theodore Barry and Associates, Inc. (1971). *Industrial engineering study of hazards associated with underground coal mine production.* Los Angles, California; Author, NTIS No. PB 207 226.

Theodore Barry and Associates, Inc. (1972). *Accident prediction investigation study.* Los Angles, California; Author, NTIS No. PB 221 000.

Van Zelst, R. H. (1954). The effect of age and experience upon accident rate. *Journal of Applied Psychology, 38*(5), 313–317.

Chapter 3
Types of New Employee: Experience and Pre-entry Safety Expectations

3.1 Introduction

Every person that starts a new job can be classified as a new employee. This is true, regardless of the nature of their previous employment history. Of course whether a new employee is a school leaver or has many years of previous work experience will have implications for their new employee-associated safety risks. However, previous job experience does not remove all the safety risks associated with being a new employee. In Chap. 3, different types of new employee are defined, and how safety issues vary across the four types of new employee is discussed. Arguably, an organization that understands the specific safety issues that are associated with a new employee will be in a better position to manage that employee's safety. It is also important that new employees understand their own vulnerabilities, and strategies are discussed which allow an organization to help different types of new employee protect themselves from risk.

3.2 Job Tenure and Job Experience

Before defining different types of new employee, it is necessary to briefly discuss the distinction between cumulative job tenure and job experience. The classification of new employees into types in the next section uses the nature of job experience as the main classifying dimension. It would be easier for safety management if the classification of new employees could be made using *cumulative job tenure* as the differentiating dimension. Cumulative job tenure is the sum of all days which an individual has been employed across their working life. For example, if an individual has had three jobs, and in total worked 8 years, their cumulative job tenure is 8 years. Unfortunately, it is very risky to equate cumulative job tenure with gained experience. For example, it is not always correct to assume or conclude that an

© Springer International Publishing Switzerland 2015
C.D.B. Burt, *New Employee Safety*,
DOI 10.1007/978-3-319-18684-9_3

individual that has worked for two years has more experience than an individual that has worked for 1 year. Equally, two individuals that both have 2 years of cumulative job tenure may have a vastly different level of job experience. Section 3.4 elaborates on the reasons why cumulative job tenure and experience should not be equated.

3.3 Classification of New Employees in Types

There are four primary types of new employee, which can be labeled *the school leaver employee*, *the career transition employee*, the *occupational-focused employee*, and *the career-focused employee*. The *school leaver* generally has little or no workplace experience. The *career transition* employee has a degree of workplace experience, but in a different industry and job type to that which they are now about to enter (i.e., an individual may have previously worked in a service job and is now entering a job in construction). The *occupational-focused* employee has previous experience in the same job they are entering, but in a different industry (i.e., an individual may have worked as a fitter in a small engineering business and is now entering a job as a fitter on an oil rig). Finally, the *career-focused* employee has previous experience in the same job and industry, but in a different organization(s) (e.g., the individual has always worked as fitter in the oil exploration industry). The four different types of new employee have potentially very different implications for workplace safety and will vary considerably in their safety risk during their initial period of employment in a new job.

Added to the four primary categories of new employee defined above are two further categories—these being temporary workers (or temp workers) and contractors. In the case of both temp workers and contractors, the organization does not necessarily know which of the four primary new employee categories the employee belongs to. Both temp workers and contractors are likely to be primarily employed by either the agency that provided the temp worker or the contracting organization. This situation creates a dilemma for organizations as it reduces their ability to predict the safety risk level associated either with a temp worker or with a contractor. A necessary step in managing new employee safety is to recognize that new employees will vary greatly in terms of risk and that the variation is partly based on factors associated with experience and safety expectations. These factors are discussed further in the following sections.

3.4 Defining Experience

To understand how the different types of new employee vary in terms of safety risks, it is necessary to explain and discuss what is meant by the nature of experience. Surprisingly, little research appears to have directly addressed the question, *what is experience?* A notable exception is the paper by Tesluk and Jacobs (1998). Perhaps this is because the term experience, or label experienced, seems self-explanatory. However, experience is a complex concept. Furthermore, given that a number of studies have linked high accident rates in new employees to their lack of experience, and experience forms the basis of the new employee classification described above, it is important to fully understand what is meant by experience.

Perhaps a useful place to begin is to make a distinction between experiencing something and having experience or being experienced. This distinction points to frequency and time as being important aspects of experience. Experiencing how a team works, or how equipment is operated, can be acquired on a single occasion. However, this single exposure does not make the individual experienced in the team's operational characteristics, nor does it make the individual an experienced operator of the equipment. Furthermore, the difference between experiencing something and having experience has important implications for employees' understanding of how effective prestart training can be at reducing new employee accidents, and perception of the effectiveness of prestart training is addressed in more detail in Chap. 6.

Given that experience has the characteristics of frequency, its time dimension is perhaps best defined by duration. At this point, it is also necessary to come back to the idea of cumulative job tenure, which in effect is the individual's total duration of exposure to work. Here, it is tempting to simply ask the question of how long, how many days, weeks, months, and years, does it take to become an experienced employee? At some point in time (in their job tenure), the new employee will achieve sufficient experience to allow the risk of them having an accident to decline significantly. Studies that have examined the link between experience and accidents often reach conclusions such as workers with less than one year of experience (what they really mean is job tenure) have more accidents than more experienced workers (e.g., Groves et al. 2007; Jeong 1998). This of course does not define the duration needed to become experienced. Rather, it is a grouping classification applied to a data set by the researcher.

Variation in activity during job tenure is sure to change the time required to be classified as experienced. For example, Van Zelst (1954) in a longitudinal study of America copper plant workers found a general leveling-off of accident rates after 5 months of job tenure and concluded that this particular sample of workers, in this particular organization, appears to have gained enough experience after 5 months to avoid accidents. It is also worth noting that in this organization, there was no formal prestart training. In an examination of another cohort of workers who were provided prestart training in work procedures and safety methods, Van Zelst (1954) found that the accident rate declined after 3 months of tenure. Thus, length of time or job tenure is likely to be a poor indicator of experience and a poor predictor of new employee safety. The important factor is what has been done during a person's employment tenure and how these activities map onto what is required in their new job.

While elapsed cumulative time or duration of job tenure is a very convenient variable to measure, organizations must not assume that once a specific period of work has passed, that worker can be considered experienced. The time frame or duration to become experienced is likely to be variable and related to variability in what the employee is doing on the job (discussed further below), and to the nature of the employees cumulative job tenure (what they have done in the past). This is not to say that duration of job tenure is not important. On the contrary, it can be important in that it provides for the possibility of *variability* and *similarity* of experience, two factors which are essential for an individual to become experienced.

Variability and similarity of experience complicate the frequency and time components of experience. As such, variability and similarity have a major influence on the ability of an employee's experience to generalize from one situation to another situation and on whether an individual can be classified as experienced or not. *Variability* is defined as the variation in previous experience. For example, an operator of an earthmover may have gained all their previous experience in a single make and model of earthmover, or they may have experience in operating several different makes and models of earthmover. Equally important is the operator's experience in different operating environments, which could range from a single type of terrain to many different terrains. Added to this is variability in other factors such as work pressure, safety climate, equipment maintenance schedules, and co-workers.

Similarity is defined as the relationship between the variability of experience from previous employment and the operating characteristics of the new job. Greater similarity should improve the generalization of previous experience to the new job. That is, make it more likely that the new employee can be considered as experienced or at least become experienced faster. Consider again the example of the operator of earthmoving equipment. A new employee may have an employment record that indicates that they have 10 years of cumulative job tenure as an operator of earthmoving equipment. However, the key questions are as follows: Was any of that experience gained in an operating environment which is the same as, or similar to, that which exists in the new job? Was the equipment operated in the previous jobs the same as the equipment being used in the new job? If the answer to these two questions is no, then the new employee cannot be considered an experienced operator in terms of their new job. In contrast, the more variability the individual has in their previous experience, the more likely the answers to these questions will be yes.

In support of the limited ability of experience to generalize is a key finding from Leigh's (1986) analysis of the relationship between accidents and individual and job characteristics. Using a logistic regression approach, they found that 'general work experience' had no significant relationship with accidents. While it is sometimes hard to know how to interpret non-significant results, this finding is consistent with the limited generalizability of experience. It is also worth noting that the mean work experience in their sample ($N = 4962$) was 16.037 years (standard deviation = 12.219). Thus, participants in the sample had on average a significant amount of general work experience, and within the sample, there was substantial variance in this measure. Yet this vast previous experience was not associated with a reduction in accidents. That is the participants' previous experience appears to have had very little generalization to their current employment situation in terms of protection from accidents.

In summary, perhaps the first key step in managing new employee safety risks is to understand their level of experience or whether they can be classified as experienced. Of course, one might argue that it is logically impossible to hire an experienced worker. Rather, an organization can only ever employee an individual that has experience in similar work. Using the word similar is more advisable than identical because it is logically impossible for a past job to be identical to a new job. If we just consider the co-workers: for two jobs to be identical, all individuals who worked with the individual in their old job would also have to move to the new job. However, while the later statement is logically true, employers can expect a degree of transfer of past experience to a new job. The nature and degree of experience transfer is discussed below and will vary depending on which type of new employee is hired. Without a careful consideration of experience, an organization can make an incorrect assumption about a new employee. For example the organization might incorrectly assume that a new employee with a previous work history can be considered experienced, and based on this assumption they need less supervision or training.

3.4.1 The Transfer of Experience

At this point, I will not explore in detail the issues around the acquisition of experience in a specific situation. These issues are dealt with in Chap. 7 where concepts such as familiarization and co-worker compensatory behaviors are discussed in detail. For now, it is sufficient to realize that new employees vary in experience and this variance has important implications for safety. Table 3.1 shows the key variables which an organization should consider: the type of new employee, the experience they have before starting their new job, the ability of that experience to generalize to the new job, and the relative time it may take for the new employee to be considered experienced. It is impossible to actually place specific times into the last column of Table 3.1, as this will vary considerably from job to job, from organization to organization, and from individual to individual. However, the general pattern shown in Table 3.1 is likely to apply to most situations.

Further complicating the ability to specify how long an individual may take to become experienced is the general finding that as workers age, their ability to quickly learn new tasks, gain new skills, gain knowledge, and therefore become an experienced employee may decline. Of course this is a gross generalization, but research does support an increase in the effort required to reach a specified level of mastery as age increases (e.g., Gist et al. 2006). Thus, in addition to which new employee category that an individual belongs to (i.e., career transition, occupational focused, or career focused), the individuals' age may have an impact on the speed at which they can become experienced. Organizations may need to make training and supervision allowances to ensure that older workers have the time necessary to acquire experience.

Table 3.1 Parameters of experience and time to become an experienced employee in a new job for each new employee category

New employee category	Level of relevant entry experience	Similarity of previous experience	Variability of previous experience	Generalization of previous experience to new job	Time to become experienced in new job	New employee risk level
School leaver	Nil	–	–	–	Considerable	Extreme
Career transition	Nil	–	–	–	Considerable	Extreme
Occupational focused	Some	Yes	Yes	Yes	Some time required	Moderate
		No	No	No	More time required	High
Career focused	Most	Yes	Yes	Yes	Quickest, but still some time required	Moderate/ low
		No	No	No	More time required	High

The right-hand column of Table 3.1 provides a new employee risk estimate for each type of new employee (although it makes no allowance for age). The risk estimate uses a simple *extreme* to *low* scale, but should serve to illustrate that not all new employees are the same in terms of how their past experience, as defined by their new employee category, will influence their safety risk in a new job. Furthermore, note that no type of new employee is fully protected from safety risks by their past experience (past employment history). It is also important to note that experience is only one factor which influences new employee safety risk. Thus, while a new employee may be classified as a moderate-to-low risk in terms of Table 3.1, there may be other factors associated with their entry into a new job which increase their accident potential. Section 3.7.2 in this chapter discusses the assessment of job applicant experience at the time of recruitment and how this can be used to help ensure new employee safety.

3.5 New Employee Safety Expectations

All four types of new employee will also vary in their *safety expectations*, ranging from their new job that has virtually no safety risks, through to the job is inherently very dangerous. These safety expectations form part of new employee's psychological contract with the employer (Burt et al. 2012; Sully 2001; Walker and Hutton 2006). Psychological contract theory is based on the perceived obligations (expected behavior) between employees and their employer and has social exchange theory (Blau 1964) and the concept of reciprocity (Gouldner 1960), as central components. McLean et al. (1998, p. 697) define the psychological contract as 'the idiosyncratic set of reciprocal expectations held by employees concerning their obligations (what they will do for the employer) and their entitlements (what they expect to receive in return).' If new employee's safety expectations are unrealistic, they are likely to be exposed to unexpected risk in their new job.

Safety-specific expectations are likely to vary across the four types of new employee in the following way. The *school leaver employee* is likely to have the most idealistic (unrealistic) safety expectations, the *career transition employee* perhaps has a slightly more realistic view of workplace safety, and finally, the *occupational-focused and career-focused* new employees are likely to have somewhat more realistic expectations of the safety realities of the type of work they are applying for and for how organizations manage safety. Research by Weyman and Clarke (2003) has shown how expectations about safety risks can vary considerably across different personnel groups. It is also the case that expectations based on experiences in one organization do not necessarily transfer to another. Organizations vary greatly in terms of their safety culture and overall commitment to employee safety, and this variation can conflict with new employee's experience-based safety expectations.

The importance of understanding new employee safety expectations is further reinforced if expectations are considered within the framework of *risk homeostasis theory*. Wilde and colleagues developed risk homeostasis theory (see Wilde et al. 2002; Simonet and Wilde 1997) which proposes that as safety features (expected or real) are added to a system, users tend to increase their exposure to risk because they feel better protected. For example, if a new employee expects that equipment is well maintained, they may use equipment without checking its functionality. Similarly, if a new employee expects that co-workers will remove hazards from the workplace, or not create hazards, they may not actively engage in as much monitoring for hazards. Put it another way, incorrect safety expectations can lead a new employee to take unexpected risks.

Despite the importance of safety expectations, there appears to have been relatively little research on the topic. Burt et al. (2012) attempted to address this gap in the literature with two studies. Study 1 collected data on safety expectations from school students who were about to transition to work. The general pattern across the students sampled was to hold very high expectations about how both management and co-workers would ensure their safety when they entered the workplace. Furthermore, they also gave relatively large ratings of trust in both management and co-workers to ensure their safety. Study 2 obtained safety expectation data from 40 school leavers (students leaving school and entering the workforce) and matched it with safety data from the manager/supervisor of the job they were about to enter. Analysis of the matched data showed that the school leavers significantly underestimated job risk and significantly overestimated co-worker safety behavior and co-worker safety reactions toward new employees. Collectively, the two studies conducted by Burt et al. (2012) clearly show a tendency for school leaver job applicants to have unrealistic and dangerous safety expectations. Holding unrealistic safety expectation could easily lead to a new employee getting into a situation where an accident results. Thus, unrealistic safety expectations may be one of the factors which are associated with new employees having more accidents.

3.5.1 Experience with Accidents and Expectations

Related to the issue of safety expectations is an individual's experience with accidents. While organizations do not want individuals to directly or indirectly experience an accident, this clearly happens. Furthermore, over an individual's life, it would be very unlikely that they do not at least indirectly experience some form of accident. Certainly, everyone is from time to time exposed to media reports of significant accidents. Research on the impact of experience with workplace accidents does suggest that it has a positive impact on subsequent safety behavior. For example, studies by Laughery and Vaubel (1989), and Kouabenan (2002) both found positive correlations between safety behavior and accident experience, suggesting that individuals become more cautious if they have an accident experience. This might also be interpreted as the individual becomes more realistic in their safety-specific expectations.

Laughery and Vaubel (1989) and Kouabenan's (2002) research raises an interesting dilemma for organizations. Should they reject job applicants that have a history of accidents on the basis that they may belong to the group often labeled 'accident prone' (see Chap. 5, Sect. 5.5.5), or is there some advantage to be gained from employing individuals who have previously experienced (directly or indirectly) one or more workplace accidents? The later individuals may well have a healthy respect for workplace risk, have realistic safety expectations, and be more inclined to be safety compliant and participative. There really is no clear answer to this question. What is clear is that safety expectations are a key factor in new employee safety.

3.6 Experience and Expectations

The results from Burt et al. (2012) clearly suggest that there is a link between an individual's degree of previous experience and their safety-specific expectations. As with the general principle of having experience or being experienced, the concepts of variability and similarity are equally important for the development of realistic safety expectations. A key factor stemming from variability and similarity of previous experience is that individuals should have variable expectations for a range of different aspects of work. For example, they should have an understanding that co-workers can behave in different ways, and thus, their expectations about how co-workers will behave should reflect this variability. Similar examples can be given for the provision of safety equipment, about the maintenance of equipment and about the influence of work pressures on safety. The more the variability in an individual's previous experience, and the greater the similarity between those circumstances and the nature of that experience and the current situation (the new job), the greater the likelihood is that their expectations will be realistic. Of course, a realistic safety expectation might simply be that the unexpected is indeed to be expected.

Table 3.2 Safety-specific expectations and degree of expectation-driven risk exposure for each new employee category

New employee category	Level of relevant entry experience	Similarity of previous experience	Variability of previous experience	Safety-specific expectations	Expectation-driven risk exposure
School leaver	Nil	–	–	Unrealistic	High
Career transition	Nil	–	–	Unrealistic	High
Occupational focused	Some	Yes	Yes	More realistic	Moderate
		No	No	Less realistic	High
Career focused	Most	Yes	Yes	Most realistic	Low
		No	No	Less realistic	Moderate

Table 3.2 is an attempt to show the general nature of safety expectations across the four categories of new employee and also shows the associated level of safety expectation-driven risk exposure. The expectation-driven risk exposure level is based on the predictions of risk homeostasis theory, where less realistic expectations about safety aspects in a workplace may lead an individual to engage in behaviors which turn out to be risky, or not to engage in behaviors, such as monitoring and being careful, which will help ensure safety.

3.7 Managing the Risks

Two key risk factors (experience and expectations) associated with different types of new employee have been identified. In the remainder of this chapter, I offer suggestions of how these risks can be assessed and managed. Given the vast variability in circumstance that is likely to be associated with readers of this work, it must be noted that individuals adopting the approaches outlined below may need to tailor them to their specific circumstances. However, in the main, the general principles and processes should hold for most, if not all, organizations.

3.7.1 Assessing New Employee Experience

Assessing a job applicant's experience by examining their work history in their CV, or their entry on work history in an application blank, is likely to provide a very poor predictor of the applicant's *experience-related risk potential*. As noted above, cumulative work tenure does not equate to experience. Experience-related risk potential is defined as the degree to which the applicant's lack of relevant previous experience exposes the applicant to risk in the new job (should they be hired). Table 3.1 outlines how this varies across the four types of new employee (type of job applicant in the case of recruitment). CV and application blank information may give an indication of cumulative job tenure, but gauging experience variability and similarity is virtually impossible, based solely on cumulative job tenure.

The employee selection literature is well developed in the area of structured employment interview development (see Huffcutt 2011; Levashina et al. 2014 for useful reviews and guidelines). The addition of a set of experience-related questions into a structured employment interview will provide a much more comprehensive profile of an applicant's experience and also allow for vastly more accurate predictions of their experience-related risk potential. Furthermore, it will give an indication of how long it may take for the individual, if employed, to become an experienced operator: an indicator of the extent of supervision and training which might be needed to ensure the new employee's safety.

Table 3.3 shows experience-related questions which might be included in a structured employment interview. While job tenure has many limitations as a measure of experience, it still has the potential to add useful information to a measurement model, and as such, questions relating to tenure are included in Table 3.3. The answers to the questions are numerical, and a larger overall score (summed across all questions) can be interpreted as a greater level of experience. That is not to say a larger score indicates that the applicant can be considered experienced, just that they have more experience which should reduce the time it will take for them to become experienced in the job they are applying for. A recruiter would also be advised to carefully note when the answer to any of the questions is zero. Measuring experience does not, in its self, mitigate against the risks associated with a lack of experience. However, detailed information on experience allows the organization to put in place steps to reduce the risk, such as extra training and additional supervision. Measuring experience using more detailed questions, such as those shown in Table 3.3, will also allow an organization to inform new employees about the limits of their previous experience and its ability to keep them safe in their new job.

3.7.2 New Employee Safety Expectation Assessment and Development

Previously, in this chapter, I have described how cumulative job tenure can help develop realistic safety expectations. Clearly, the school leaver applicant has not had the opportunity to adjust their safe expectations based on previous work experience. However, educators and parents can play an important role in safety-specific expectation setting. Parents and educators can deliver key messages about workplace safety either informally as part of parenting or formally as part of the school curriculum activities. While parents may not have a legal responsibility to ensure their child's safety at work, as parents they do have a moral responsibility. School system education in occupational health and safety has been called for by a number of authors (e.g., American Public Health Association 1995; Bush and Baker

Table 3.3 Experience-related questions for use in an employment interview

Interview Questions	Scoring
In total how many different jobs have you had?	……….
In total how long (in months) have you worked for?	……….
How many different organizations have you work for?	……….
Thinking about the job you are applying for	
How many different organizations have you undertaken this job in?	……….
How many different work groups or teams have you performed this job with?	……….
How many different work environments have you undertaken this job in?	……….
How many different makes or models of equipment have you operated?	……….
Considering your total working life, what percentage includedtasks similar to those of the job you are applying for?	……….
How similar is the work environment of the job you are applying for to work environments you have workedin the past?	
1………..2………3………4………5………6………7	……….
Not worked in this type of environment Extremely similar	
How similar is the equipment used on the job you are applying for compared to equipment you have worked with in the past?	
1………..2………3………4………5………6………7	……….
Not worked with this type of equipment Extremely similar	
Overall, how similar is the job you are applying for to the work you have done in the past?	……….
1………..2………3………4………5………6………7	
Not done this type of work before Extremely similar	
Thinking about your training and education for the job you are applying for	
How many months of job related education have you received?	……….
How many training programs relevant to the job you are applying for have you attended?	……….
How many on-the-job-mentors have you worked with to develop your job skills and knowledge?	……….
Total Score	……………..

1994; Castillo 1999; Schulte et al. 2005). However, specific education at the high school level in workplace risks, hazards, and safety is arguably at best inconsistent across counties (Sas 2009). On a positive note, research has shown that such education can increase knowledge about safety (e.g., Lerman et al. 1998; Linker et al. 2005). However, on a cautionary note, there is some evidence that its efficacy is likely to be determined by a number of factors associated with the instructor or teacher (Pisaniello et al. 2013).

From an organizations' perspective, safety expectation assessment and development might be undertaken by using a three-stage process: Stage 1: documentation via the job vacancy advertisement and the job description material; Stage 2: selection measures with an emphasis on assessing the applicant's safety expectation foundation and status; and Stage 3: on-boarding with a focus on building job-specific realistic safety expectations. Stage 1 ensures that all job applicants are provided with detailed safety information about the job they are applying for. Stage 2 provides for the assessment of the job applicant's safety expectations. Finally, Stage 3 is applied to individuals that are selected for the job and delivers these new employees job-specific safety expectation setting information as part of the on-boarding, or induction process. Combined, the three stages should reduce new employee safety risks associated with unrealistic safety expectations. The three stages are explained in more detail below.

3.7.2.1 Stage 1: Documentation

What we expect is based on what we know. This simple premise formed the bases for the development of realistic job preview theory. Realistic job previews provide a clear understanding of what a job entails and what competencies are required to do the job. If realistic job preview information is delivered as part of a recruitment process via inclusion in the job advertisement (see Chap. 5, Sect. 5.4), and in the associated job description and person specification documents, the information allows for individuals to select themselves in or out of an applicant process. Those that stay in the process are likely to see the organization as more trustworthy, are likely to have developed coping strategies for aspects of the job, are likely to be more satisfied and committed once working in the job, and are likely to remain in the job for longer. Realistic job previews are really about setting realistic expectations or lowering expectations to a realistic level (e.g., Buckley et al. 1998; Morse and Popovich 2009).

The realistic job preview literature does not deal specifically with safety, although safety is clearly an aspect of a job which a job applicant, and later a new employee, needs to have very realistic expectations about. Safety-specific expectation knowledge might be described as knowledge of the job's overall risk level, knowledge of the job's context (environment) risk level, knowledge of the job's equipment risk level, knowledge of the job's co-worker experience, and knowledge of the job's performance demands (physical and cognitive load). The safety-specific realistic job preview should not include any information about the organizations' attempts to mitigate the job's risks. For example, it should not say 'The organization operates a state of the art safety management system.' While this may be true, such a statement is likely to reduce the applicants' (new employees) risk perceptions (make them potentially unrealistic), and according to *risk homeostasis theory* (Wilde et al. 2002; Simonet and Wilde 1997), perceived safe guards will increase risk taking. Further issues associated with how employee's perceptions of organizational processes can influence risk taking are explored in detail in Chaps. 5 and 6.

In my view, realistic safety information should form a specific section in a job description. Figure 3.1 is an example of what a safety section in a job description might look like. The actual information in the safety section of a job description should be derived from a careful analysis of the job, from internal accident data associated with the job, and from global accident trends associated with the job. For the information to help with expectation setting, it needs to be as realistic (accurate) as possible. Furthermore, the organization should not distort or adjust the ratings based on their expectations of how safety management systems are likely to reduce the risks.

While it seems relatively straight forward, and pretty much common sense, to include information about safety risks and hazards in a job description, it is unclear whether this really is common practice. A study by Ramsay et al. (2006) examined the job descriptions of nurses from 29 Veterans Affairs hospitals from across the USA. The job descriptions were examined by an expert panel of occupational nurses for the degree to which they incorporated the 12 primary occupational safety and health risk factors which emergency department nurses face, as defined by Occupational Safety and Health Administration (OSHA). Surprisingly, virtually, none of the position descriptions included the key safety information. This seems like a *missed opportunity*, particularly given the importance of taking all possible steps to alert new employees to safety risks and hazards.

Safety Specific Realistic Job Preview Information

The information in this section describes the degree of safety risk associated with aspects of this job.

It is important for the job incumbent to have realistic expectations of the risks and safety issues associated with this job.

Overall job safety risk level - **extreme**

Global accident risk for this job - **high**

Safety risk level associated with working in job environment - **high**

Safety risk level associated with working with job machinery - **high**

Co-worker job experience level – Minimum?…….., Maximum?…….., Mean?…….

Co-worker risk level - ?…………

Job's physical demands - **high**

Job's cognitive demands - **moderate**

Fig. 3.1 Example of tree harvester (forester) safety-specific realistic job preview information for use in a job description

The example of a safety profile for use in a job description shown in Fig. 3.1 includes the job's overall risk level, the job's environment risk level, and the job's equipment risk level on a 4-point scale: minimal, moderate, high, and extreme. Determining which scale point to select can be a somewhat subjective process, for example, the job of tree harvester is generally extremely risky, is performed in an highly risky environment and uses highly dangerous equipment. Co-worker job experience can be described somewhat more objectively by using employee's job tenure data (minimum, maximum, and mean) associated with the job (remembering that new employees are more risky to work with than experienced employees). If the job's co-workers on average have short job tenure, it is more likely that the new employee will be working with less experienced co-workers, and the co-worker risk level will be higher. Finally, the physical and cognitive demand level can be determined by examining what tasks are involved in the job.

3.7.2.2 Stage 2: Assessment

The primary method which could be used to assess job applicant's safety expectations is a number of questions presented in an employment interview. Table 3.4 shows questions which could be used. It is also possible to use other questions, such as *What is the safety risk level associated with the job you are applying for?* and *Will it be necessary for you to be vigilant about safety in the job you are applying for?* However, such questions are very open to socially desirable responses. In contrast, the questions shown in Table 3.4 should provide a more objective measure of factors which should positively influence realistic safety expectations. Each question will generate a numeric response which can be totaled. A larger overall score should be indicative of a higher probability that the individual will have realistic safety expectations. Recruiters would also be advised to note when an

Table 3.4 Safety expectation interview questions

Interview questions	Score
How many jobs have you worked in where safety was a concern?	....
How many conversations have you had with colleagues, supervisors, parents, friends, etc. about workplace safety issues?	....
How many hours of instruction on workplace safety did you receive at school?	....
How many hours of safety training have you received?	....
How many workplace accidents have you had?	....
How many workplace accidents have you seen others have?	....
Have you previously worked in the job you are applying for? (yes = 1 point, no = 0)	....
Have you read the safety-specific risk section in the job description? (yes = 1 point, no = 0)	....
Total score (Higher score suggests potential for more realistic safety expectations.)	....

applicant achieves a zero score for any question, as this may red-flag a potential safety expectation issue.

3.7.2.3 Stage 3: Induction

To help set realistic safety expectations, an organization should use a *realistic safety preview* as part of the induction process for new employees. This would be particularly important for new employees that score low on the safety expectation assessment outlined in Sect. 3.7.2.2. New employee safety risks associated with their safety expectations should be able to be decreased or removed through organizational socialization/induction processes (see Bauer et al. 2007), or an *on-boarding* strategy (see Bauer and Berrin 2011), which attempts to align their safety expectations with the *safety realities* of the workplace. Safety issues associated with socialization processes are also discussed in Chap. 6. This attitudinal alignment of safety expectations with the safety reality of the workplace would be in addition to prestart training in equipment use and safety procedures which an organization might offer.

Expectation alignment might be achieved by a simple data collection and feedback process. The scales used in Burt et al. (2012) (*Expected management safety behavior*, *Expected co-worker safety behavior*, *Expected worker reactions to new employees*, and *Expected safety behavior by new employees,* see Chap. 9 for scale items) could be administered to supervisors and job incumbents on an annual basis. Job incumbents should also be asked to rate job risk (e.g., please indicate your expectation of the **safety risk associated with the job** by placing a mark on this 100-point scale where 0 = not at all risky and 100 = extremely risky). During new employee induction, the new employee would be given the same questions (using wording in first person singular future tense—see Chap. 9) to complete. After the new employee has completed the expectation scales, they would be scored and responses entered onto a feedback sheet such as that shown in Table 3.5.

Table 3.5 Feedback form for use during the safety expectation alignment component of new employee induction

Expectation scale	New employee scale score	Mean job incumbent scale score	Job incumbent minus new employee scale score[a]
Expected co-worker safety behavior			
Expected management safety behavior			
Expected worker safety reactions to new employees			
New employee safety-related behavior			
Job safety risk rating			

[a]Positive scores in this column suggest unrealistic safety expectations

Discrepancy between the new employee's responses and the job incumbents' responses, particularly positive scores in the last column of Table 3.5, should be considered as possible evidence of unrealistic safety expectations. Where such evidence is found, the new employee needs to be made aware that their perceptions are out of line with current employees. They should also be made aware that their distorted perceptions, and their unrealistic safety expectations, could place them and/or their co-workers at risk of an accident.

3.8 Conclusions

It is clear that organizations can have a range of different applicants applying for jobs. The four categories of job applicant, and thus the four types of new employee, have associated with them different safety risks due to experience and expectation factors. These factors have been outlined in this chapter. It is clear that organizations can make incorrect assumptions about the value of past experience and how it might help ensure new employee's safety. This chapter offers suggestions on how experience can be measured in a relatively complex way and how this information can used to help ensure new employee safety. While most organizations probably give at least some consideration to job applicant's experience, few probably consider safety expectations.

Safety expectations are relatively easy to deal with, yet extremely dangerous to ignore. While we currently do not know how many accidents can be attributed to unrealistic safety expectations, it is likely to a significant number. The issue lies in the fact that new employees are likely to overestimate the degree to which the organization and their co-workers can, and will, protect their safety. Furthermore, we know that perceptions that a system has protective components are associated with enhanced risk taking. This chapter offers a relatively simple three-stage process by which safety expectations can be dealt with. In the absence of such a strategy, it is to be expected that new employees will exposure themselves to safety risks which they were not expecting.

References

American Public Health Association. (1995). Protection of child and adolescent workers—Policy statement adopted by the governing council of the American Public Health Association, November 2, 1994. *American Journal of Public Health, 85*, 440–442.
Bauer, T. N., & Berrin, E. (2011). Organizational socialization: the effective onboarding of new employees. In S. Zedeck (Ed.), *APA handbook of industrial and organizational psychology, Vol 3: Maintaining, expanding, and contracting the organization* (pp. 51–64). Washington, DC: American Psychological Association.

Bauer, T. N., Bodner, T., & Tucker, J. S. (2007). Newcomer adjustment during organizational socialization: A meta-analysis review of antecedents, outcomes, and methods. *Journal of Applied Psychology, 92*(3), 707–721.

Blau, P. (1964). *Exchange and power in social life.* New York: Wiley.

Buckley, M. R., Fedor, D. B., Veres, J. G., Wiese, D. S., & Carraher, S. M. (1998). Investigating newcomer expectations and job-related outcomes. *Journal of Applied Psychology, 83*, 452–461.

Burt, C. D. B., Williams, S., & Wallis, D. (2012). New recruit safety expectations: Relationships with trust and perceived job risk. *Safety Science, 50*, 1079–1084.

Bush, D., & Baker, R. (1994). *Young workers at risk: Health and safety education and the schools.* Berkeley, CA: University of California, Berkeley.

Castillo, D. N. (1999). Occupational safety and health in young people. In J. Barling & E. K. Kelloway (Eds.), *Young workers: Varieties of experience* (pp. 159–200). Washington, DC, US: American Psychological Association.

Gist, M., Rosen, B., & Schwoerer, C. (2006). The influence of training method and trainee age on the acquisition of computer skills. *Personnel Psychology, 41*, 255–265.

Gouldner, A. W. (1960). The norm of reciprocity: A preliminary statement. *American Sociological Review, 25*, 161–178.

Groves, W. A., Kecojevic, V. J., & Komljenovic, D. (2007). Analysis of fatalities and injuries involving mining equipment. *Journal of Safety Research, 38*, 461–470.

Huffcutt, A. I. (2011). An empirical review of the employment interview construct literature. *International Journal of Selection and Assessment, 19*(1), 62–81.

Jeong, B. Y. (1998). Occupational deaths and injuries in the construction industry. *Applied Ergonomics, 29*(5), 355–360.

Kouabenan, D. (2002). Occupation, driving experience, and risk and accident perception. *Journal of Risk Research, 5*(1), 49–68.

Laughery, K. R., & Vaubel, K. P. (1989). The role of accident experience on subsequent accident events. In A. M. Feyer & A. Williamson (Eds.), *Occupational injury: Risk, prevention and intervention* (pp. 33–43). London: Taylor and Francis.

Leigh, J. P. (1986). Individual and job characteristics as predictors of industrial accidents. *Accident Analysis and Prevention, 18*(3), 209–216.

Lerman, Y., Feldman, Y., Shnaps, R., Kushnir, T., & Ribak, J. (1998). Evaluation of an occupational health education program among 11th grade students. *American Journal of Industrial Medicine, 34*, 607–613.

Levashina, J., Hartwell, C. J., Morgenson, F. P., & Campion, M. A. (2014). The structured employment interview: Narrative and quantitative review of the research literature. *Personnel Psychology, 67*(1), 241–293.

Linker, D., Miller, M. E., Freeman, K. S., & Burbacher, T. (2005). Health and safety awareness for working teens: Developing a successful statewide program for education teen workers. *Family and Community Health, 28*, 225.

McLean, P. J., Kidder, D. L., & Gallagher, D. G. (1998). Fitting square pegs into round holes: Mapping the domain of contingent work arrangements onto the psychological contract. *Journal of Orgainzational Behavior, 19*, 697–730.

Morse, B. J., & Popovich, P. M. (2009). Realistic recruitment practices in organizations: The potential of generalized expectancy calibration. *Human Resource Management Review, 19*, 1–8.

Pisaniello, D. L., Stewart, S. K., Jahan, N., Pisaniello, S. L., Winefield, H., & Braunack-Mayer, A. (2013). The role of high schools in introductory occupational safety education—Teacher perspective on effectiveness. *Safety Science, 55*, 53–61.

Ramsay, J., Denny, F., Szirotnyak, K., Thomas, J., Corneliuson, E., & Paxton, K. L. (2006). Identifying nursing hazards in the emergency department: A new approach to nursing job hazard analysis. *Journal of Safety Research, 37*, 63–74.

Sas, K. (2009). *OSH in the school curriculum: Requirements and activities in the EU member states*. European Agency for Safety and Health at Work. European Communities. https://osha.europa.eu/en/publications/reports/TE3008521ENC. Accessed 4 April 2014.

Schulte, P. A., Stephenson, C. M., Okun, A. H., Palassis, J., & Biddle, E. (2005). Integrating occupational safety and health information into vocational and technical education and other workforce preparation programs. *American Journal of Public Health, 95*, 404.

Simonet, S., & Wilde, G. J. S. (1997). Risk: Perception, acceptance and homeostasis. *Applied Psychology: An International Review, 46*, 235–252.

Sully, M. (2001). *When rules are not enough: Safety regulation and safety culture in the commercial driving context*. Paper presented at the Insurance commission of Western Australia Road Safety Conference, Perth, WA.

Tesluk, P. E., & Jacobs, R. R. (1998). Towards an integrated model of work experience. *Personnel Psychology, 51*, 321–355.

Van Zelst, R. H. (1954). The effect of age and expereince upon accident rate. *Journal of Applied Psychology, 38*(5), 313–317.

Walker, A., & Hutton, D. M. (2006). The application of the psychological contract to workplace safety. *Journal of Safety Research, 37*, 433–441.

Weyman, A. K., & Clarke, D. D. (2003). Investigating the influence of organizational role on perceptions of risk in deep coal mines. *Journal of Applied Psychology, 88*(3), 404–412.

Wilde, G. J. S., Robertson, L. S., & Pless, I. B. (2002). For and against: Does risk homoeostasis theory have implications for road safety. *British Medical Journal, 324*, 1149–1152.

Chapter 4
The Job's Safety Risk Profile

4.1 Introduction

To manage an employee's safety, the employee and the organization must fully understand the safety issues and hazards which are associated with their job. For many jobs, the safety issues are very well defined, employees are knowledgeable about these safety issues, and expect them. Further, strategies are often in place to minimize the possibility that known safety issues and hazards will result in an accidents. For example, specific training is provided, specific safety equipment is used, and specific procedures are used. Unfortunately, it is not too difficult to add other safety issues and hazards to a job. These fall into the unexpected category, do not need to be there, and can be controlled and removed. These additional risk and hazard components include faulty equipment, inappropriate task assignment, workload pressure, working hours and scheduling, environmental variance, co-worker behavior, and variation in supervision. Each of these factors is discussed in this chapter, with a particular focus on how they can add to the safety risks which a new employee is exposed to. Strategies to manage these additional risk factors are suggested.

4.2 Job's Safety Risk Profile

Figure 4.1 illustrates the nature of a job's safety risk profile, and what an employee can potentially encounter when they begin a new job. Education, experience, and training can prepare a new employee for normal and known safety risks. However, a number of other factors can add safety risks to a job. Figure 4.1 shows a number of these factors: equipment safety issues, task assignment, workload and performance requirements, scheduling and work hours, environmental variance, co-worker behavior, supervision, and employee silence and safety voice. These

© Springer International Publishing Switzerland 2015
C.D.B. Burt, *New Employee Safety*,
DOI 10.1007/978-3-319-18684-9_4

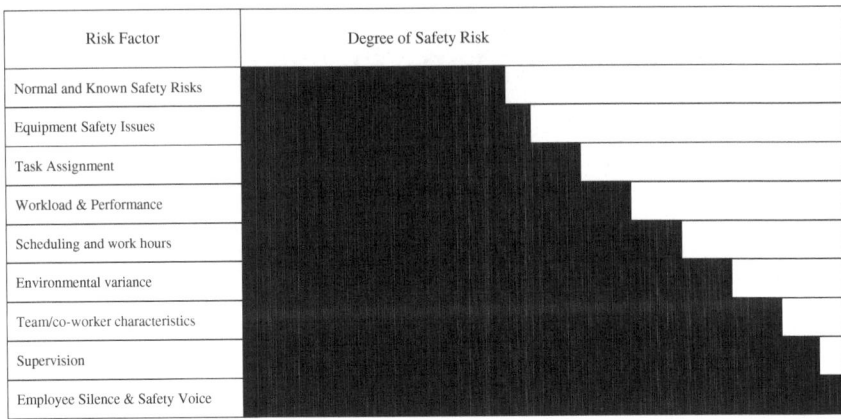

Risk Factor	Degree of Safety Risk
Normal and Known Safety Risks	
Equipment Safety Issues	
Task Assignment	
Workload & Performance	
Scheduling and work hours	
Environmental variance	
Team/co-worker characteristics	
Supervision	
Employee Silence & Safety Voice	

Fig. 4.1 Factors that can contribute to a job's safety risk profile

factors serve to illustrate how a job's safety risk profile can extend beyond normal and known limits. Each factor is more or less controllable by the organization. While the factors are discussed as separate issues, there clearly are interrelationships between them, and these are noted where appropriate.

Figure 4.1 also illustrates the additive nature of a job's safety risks. That is, added safety risks associated with each factor cumulatively increase the job's overall safety risk. The additive nature of risk is well captured by Reason's (1990) *Swiss cheese safety model*. Each factor represents a risk layer, and the probability of an accident increases as the number of risk layers increases. Where possible a job which a new employee is entering should be striped of as many risk layers as possible. As will be discussed below, more senior employees, who are experienced, are likely to be better able to cope with job risks which are difficult or impossible to remove.

4.2.1 Normal and Known Safety Risks

Arguably, the first time a new employee undertakes an activity, the activities normal and known safe risks, might increase an individual's safety risk simply because the new employee has never undertaken the activity before. For example, an individual piloting a boat into a port for the first few times is probably exposed to more risk from rocks at the harbor entrance, compared to the risk exposure of subsequent visits to the harbor. Chapter 3 explains in detail the nature of experience, its distinction from the label 'experienced,' and the limited ability of experience to transfer from one situation (job) to another. Therefore, it is very important for organizations to realize that even the normal and expected safety risks associated with a job will initially have heightened risk for a new employee. The new employee's previous experience (if they have any) will not completely eliminate this heightened risk. In order to allow the new employee to learn how to deal with a jobs normal and expected safety risks, all other risks, such as those discussed below, should be removed or controlled.

4.2.2 Equipment Safety Issues

A number of studies have identified a relationship between equipment and accidents. For example, Driscoll et al. (1995) identified equipment problems as a contributing factor in their analysis of forestry and sawmill fatalities in Australia. There are perhaps four key aspects of equipment to be considered in relation to safety: operational risk, design, age, and maintenance. Each of these aspects is discussed below in relation to new employee safety.

Equipment's *operational risk* refers to the risk associated with how a piece of equipment functions in order to achieve its purpose or goal. As a simple example, the task of felling a tree can be performed with a number of different pieces of equipment: a hand saw, an axe, a chainsaw, or a mechanical harvester. Of the 4 equipment options, the chainsaw is perhaps the most hazardous to operate. Thus, in a work situation, all the equipment being operated should be able to be scales or ranked in terms of operational risk. Where possible a new employee, in their initial period of employment, *should be assigned to use the least risky equipment in terms of its operational risk*. While the example above relates to different equipment which can be used for the same task, in some jobs different equipment is used to perform different aspects of a job. If this equipment varies in operational risk, a new employee should initially be assigned to work on the least risky equipment. Or put a different way, the new employee should not be immediately asked to operate the most risky equipment.

Equipment design and equipment age are often linked. The newest equipment is very likely to have the latest design features, some of which are likely to reduce the safety risks associated with its operation. Equipment in many work situations is likely to vary in terms of age and thus in terms of design features. Some equipment may be new and have many inbuilt safety features, while other equipment could be reaching the end of its serviceable life. While some industries may be generally characterized by relatively new equipment, this is not the case for all industries. For example, McLaughlin and Mayhorn (2011) found that many farmers were using decades-old equipment. As equipment ages, the risks associated with operating it can increase. This is not only due to the equipment having less inbuilt safety features, but also due to the equipment becoming less responsive to operator control, and less load tolerant, as it ages. Arguably, the safe operation of older equipment can require more operator skill. As such, new employee safety can be enhanced by ensuring they operate the newest equipment that is available.

Equipment risks can be reduced through well-designed equipment maintenance programs. Indeed, the literature on equipment maintenance now identifies *risk-*

based maintenance as a key aspect of safety (e.g., Arunraj and Maiti 2007, 2010; Bertolini et al. 2009; Khan and Haddara 2003; Zhaoyang et al. 2011). Risk-based maintenance focuses on the reduction of safety risk, and the achievement of optimal production, via the targeting of maintenance using knowledge of equipment failures and accidents. In other words, risk-based maintenance begins from an understanding of each piece of equipment's failure and accident potential. If an organization undertakes risk-based maintenance analysis, the information gathered can be used to determine if a new employee might be restricted from working on the equipment in their initial period of employment. That is, new employees should not be assigned to operate equipment which has a high failure and accident rate.

The three recommendation above, that new employees are initially assigned to use equipment with the least operational risk, which is the newest available model, and has the lowest risk-based maintenance score, are likely to enhance new employee safety. Unfortunately, it is not uncommon for more senior employees to be operating the newest equipment. This may be the results of norms operating within the work team where the more senior employees 'get the best gear,' or it may be driven by organizational policy. In the latter case, the strategy may help ensure the equipment is protected from misuse and damage. However, asking the new employee to operate the older and potentially less safe equipment, which has a high degree of operational risk, has the potential to be a contributing factor to new employee accidents.

4.2.3 Assigning Tasks to New Employees

A job's tasks can vary in terms of their desirability. Furthermore, it is very likely that task desirability and task safety risk are related, with the less desirable tasks also perhaps being the more risky. How tasks are assigned within a work group has important implications for new employee safety. Tasks can be formally assigned by a supervisor, alternatively, or perhaps even as well, tasks can be informally assigned by co-workers. If new employees are assigned undesirable tasks because other more senior workers do not want to do them, they are potentially being placed into a risky situation when they are in fact the least prepared employee to perform the risky task.

Task assignment is a factor which organizations can have control over. While more senior employees may not wish to undertake undesirable tasks, allowing these tasks to be assigned to a new employee is likely to be adding risk to the new employee's work. It is important to ensure that group norms have not developed which operate to informally place new employees into undesirable and potentially risky tasks. While these tasks may be a necessary part of the operation, it is not in the best interests of safety to routinely assign the new employee to perform them.

4.2.4 Assigning New Employee Working Hours and Shifts

There is clear research evidence that the likelihood of an accident increases as the number of hours worked increases (e.g., Dembe et al. 2005; Hänecke et al. 1998; Lombardi et al. 2010; Nachreiner 2001; Salminen 2010). As work hours increase, the employee's fatigue level is likely to increase (e.g., Dorrian et al. 2011), leading to problems with concentration and functional effectiveness. Folkard and Tucker (2003) report a near twofold increase in accident likelihood between 8-h and 10-h shifts, and Rosa (1995) found a 3-fold increase in accidents after 16 h of work.

Hours of work are a factor which organizations have control over. Admittedly, there will often be organizational issues which have resulted in long hour shifts (10 or 12 h) being put in place. Furthermore, it may be difficult to initially have new employees working shorter (8 h max) shifts. But, from a new employee safety perspective, restricting the initial shift duration to no more than 8 h could reduce new employee accidents. It may be the case that new employees are restricted to an 8-h shift for at least their first 3 months of employment. Certainly, if an 8-h shift is operating and overtime is available, new employees should be restricted from undertaking it for at least the first 3 months of employment.

4.2.5 Workload and Performance

Every job has some form of performance expectation. While a performance expectation or performance goal can be motivating for an employee, and clearly has advantages for the organization, initially the performance expectations need to be balanced against the increase in safety risk which may be associated with them. Performance expectations create work pressure, for example to achieve a specified degree of output in a certain time. Thus, performance expectations are closely related to workload, although the relationship is not exclusive.

Recent reviews have shown that work pressure is a factor associated with safety (e.g., Christian et al. 2009; Clarke 2006), as is high physical workload (e.g., Holcroft and Punnett 2009). Research has clearly shown that pressure to perform can increase the likelihood that an employee will take a shortcut (e.g., Hofmann et al. 1995; Wright 1986) which may result in a safety violation. Shortcuts and rule bending may allow tasks to be completed more quickly (Slappendal et al. 1993), although clearly with an associated increase in safety risk. Adding pressure to perform in the initial stages of a new employee's work is likely to increase the chances that they will adopt shortcuts and rule bending as a strategy to meet the work requirements. A period of initial employment associated with a relaxing of performance expectations may help the new employee adopt appropriate and safe work behaviors, and avoid accidents associated with fatigue and safety violations.

4.2.6 Environmental Variance

A jobs safety risk is partly determined by the operating environment (Jaffar et al. 2011). For example, Lenné et al. (2012) examined 263 significant mining incidents in Australia and concluded that issues with the physical environment where associated with the incident in 56 % of the cases. Rosness et al. (2012) provide a useful review of the literature on the relationship between environmental conditions and safe work. The physical environment in particular (which includes weather and terrain, as well as ambient conditions such as heat, vibration, lighting, and toxins) can vary considerably in terms of the preconditions which it affords for an accident to happen. While this is not the case for all jobs, for many jobs there can be considerable environmental variance in terms of where the job is performed. Consider for example the job of operating earth moving equipment. The operating environment can vary from level ground to rough ground with a considerable degree of slope. While an experienced employee may be able to cope well (work safely) in an extreme operating environment (an environment that affords preconditions for an accident), for a new employee such an environment may simply add risk to the job. Organizations should have control over the assignment of employees to the operating environment. If work is being performed in a range of operating environments, assigning new employees to initially work in the least risky or hazardous environment should help ensure their safety.

4.2.7 Team and Co-worker Characteristics

Perhaps the most variable safety risk factor associated with a job is the people that the new employee works with. Broadly speaking, this might be referred to as the psychosocial hazard. It is well established that work groups or teams can view safety in different ways (Gillen et al. 2002), and that safety related behaviors are influenced by group norms and practices (Zohar and Tenne-Gazit 2008; Zohar 1980). A job's safety risks can extend beyond normal levels if co-workers are neither safety compliant nor safety participative, and/or have norms and values which are not consistent with safety. Co-workers can change from day to day and can have vastly different attitudes toward safety, and these attitudes can also be variable from day to day. Initially, a new employee has no way of predicting how their co-workers will behave in relation to safety or how they will behave in relation to their (the new employee's) safety. However, by developing an understanding of different teams within an organization (e.g., the within team experience level and the team's collective accident record), the organization should be able to predict the level of risk and hazard associated with assigning a new employee to different teams.

Teams also vary in their ability to function both in terms of goal achievement and safety. The literature on team mental models (see Burtscher and Manser 2012, for a useful review) points to the need for teams members to have shared and organized knowledge. Cannon-Bowers et al. (1993) distinguish between four types of mental model which a team should hold: The equipment model which relates to how equipment and systems operate; the task model which relates to procedures and tasks; the team interaction model which relates to how tasks require team member interaction; and the team model which relates to team members knowledge and skills. Teams with better (more complete) mental models operate more effectively and more safely. Team mental models have two distinct properties: similarity and accuracy (Marks et al. 2000). Accuracy refers to how accurate the information held by team members is, and similarity refers to the degree to which team members share the same knowledge (Mathieu et al. 2005). Arguably, a team which collectively has an accurate mental model (shared by all members) is a better team to place a new employee into in terms of being able to collectively guide their integration into the team in a safe way.

At this point, it is also worth noting the vast literature on safety culture and safety climate. Safety culture stems from the organization and is the top-down safety values, beliefs, and norms, while safety climate is more accurately defined as the employee's perceptions of how various aspects of the working environment impact on their safety (see Bjerkan 2010, for a more detailed discussion of the relationship between safety culture and climate, and its impact on team safety). From the point of view of this section, it is sufficient to understand that an organization's safety culture (and all that it entails) may be viewed differently by different teams. When a team collectively perceives safety in the same way as the organization (assuming a positive perception), the team might be said to have a strong or positive safety climate. Furthermore, this situation (a strong or positive safety climate within a team) should make the team a safer option for the integration of a new employee.

An organization should have a clear understanding of the mental model and safety risk profile of different teams. The process of assigning a new employee to a work team or unit should take into consideration the mental model of the team, the experience level of the team, and the safety risk profile of the team. Arguably placing a new employee into a team with a strong mental model, a high degree of team member experience and a low safety risk profile should help reduce the chances of the new employee having an accident in their initial period of employment.

4.2.8 Supervision

Supervision is a key factor which can reduce a job's safety risk. Equally, a lack of appropriate supervision has been found to be associated with safety incidents (e.g., Lenné et al. 2012). Supervision has a number of dimensions. Wiegmann and Shappell (2003) suggest that appropriate supervision should provide guidance, training, leadership, oversight, and incentives. Inadequate or inappropriate supervision may lack the above dimensions, and it may also include planned inappropriate operations, failure to correct known problems, and supervisory violation of policy or rules (Wiegmann and Shappell 2003). Supervision of a new employee can provide a guide to appropriate behaviors and a barrier to inappropriate behaviors. Without supervision, the new employee is largely free to behave in a way which they think is appropriate, and given their lack of familiarity with the job, and all its associated components (see Chap. 7 for a detailed discussion of familiarity acquisition), such behaviors could be risky.

Supervisory guidance has three dimensions: attention, proximity, and continuity (Morrongiello et al. 2008). A supervisor has to actively attend to what a new employee is doing, they need to be close enough to the new employee to stop risky behavior, and to guide appropriate behavior. Furthermore, the supervisor needs to continue this level of supervision for a considerable period of time. Of course, the responsibility for the supervision of a new employee may not only rest with the team supervisor or unit manager, it may also be formally, or informally, assigned to one or more members of the team (see Chap. 8, Sect. 8.6.5 for a discussion of the risks associated with informal help provided to new employees by co-workers). In particular, the need for continuity of supervisor may necessitate a degree of formal delegation of this responsibility to team members.

There is likely to be an interaction between the frequency of new employee entry into a work unit and the level of supervision guidance that is likely to be given to each new employee. When only one new employee arrives, a supervisor may be able to focus his or her attention on this individual for a considerable period of time. In contrast, if a week or two after a new employee arrives, another new employee arrives, the supervisor's time will be divided between the two, and so on. In the latter case, the organization may need to have a strategy to ensure that appropriate supervision can be provided to all new employees.

4.2.9 Employee Silence and Safety Voice

As discussed in Chap. 2, there can be a relationship between accident rates and employee turnover in that research has shown that poor job safety can result in employees resigning from their job (Bell and Grushecky 2006; Cree and Kelloway 1997; Kincaid 1996; Ring 2010; Viscusi 1979). Furthermore, evidence is mounting which suggests that employees may leave a job because they feel they are unable to *voice* their safety concerns, or they may leave because they feel that if they do voice safety concerns nothing will be done about them (Burt et al. 2013; Cree and Kelloway 1997; Reason 1997). A new employee entering a workplace where voicing safety concerns is not the norm can be placed at risk simply because no one is communicating information they may need know in order to be safe. For example, where employee silence is the norm (Brinsfield 2013; Milliken et al. 2003), a new employee may not be told to be careful operating a specific piece of equipment which has a fault, or which is somewhat unsafe to operate, due to its age.

If exiting employees are not voicing safety concerns, then the organization may not have the necessary knowledge required to prompt corrective action, and the new employee (even those with realistic expectations of the normal safety risk profile for the job type) may be about to enter a job with an unacceptable level of safety risk (beyond those normally associated with the type of work). Thus, new employee safety will be enhanced if a workplace has a safety voicing culture, where employees freely share safety information, and this is supported and reinforced by both management and co-workers. In contrast, a new employee that enters a workplace which has a silence culture, or has employees that want to voice about safety but feel they cannot for some reason, can be exposed to more safety risk than is necessary.

In order to help ensure that new employees do not encounter an unacceptable level of safety risk as a result of employee silence, an organization can introduce the use of a *safety-specific exit voicing* process (Burt et al. 2013). This process captures safety information from employees when they resign from a job (often the precursor to a new employee arriving). Because the safety information is captured once the employee has resigned, a number of the commonly found barriers to voicing, such as work group pressure to stay silent, and managers and supervisors directing blame at voicing employees, are removed. Research suggests that a failure to voice safety issues can result from a lack of management support, sometimes labeled a 'blame culture,' where voiced safety information is used to assign blame and take disciplinary action against those believed responsible (e.g., Clarke 1998; Probst and Estrada 2009; Webb et al. 1989). Furthermore, Withey and Cooper (1989) suggested that employees weigh up the possible benefits and costs when deciding whether or not to voice their concerns.

The *safety-specific exit voicing* process can be in the form of either an interview or a survey, although an exit survey has advantages over an interview (Feinberg and Jeppeson 2000; Giacalone et al. 1997; Gordon 2011). The process aims to collect information on safety aspects of the job which the individual has resigned from, thus the information will be directly relevant to the new employee coming into the resigned employee's job. The information collected would extend to all aspects of the job, including supervision, co-worker behavior, equipment, and training. The information would be used by the organization to make safety improving changes and/or passed to the new employee during their induction process.

Research by Burt et al. (2013) on the *safety-specific exit voicing* process indicated that a large percentage of employees that have resigned from high-risk jobs are willing to complete a safety-specific exit survey. Chapter 9 provides details on the questions which Burt et al. (2013) used in their safety-specific exit survey. It is important to note that the use of a safety-specific exit process should be in addition to the development of a safety voicing culture in a workplace. A safety voicing culture requires leadership (Conchie et al. 2012) and co-worker support and commitment (Tucker et al. 2008). The key issue is to ensure all the necessary safety information is freely shared, that safety issues which can be removed are removed, and where this is not possible all workers and particularly new employees know about these issues.

Table 4.1 Job risk profile checklist items

Risk factor	Key question	Key action
Equipment safety issues	Are new employees operating the safest available equipment?	Equipment allocation is controlled by management
	Is there an option to allow new employees to operate less risky equipment during their initial employment period?	Policy where equipment use is partly determined by employee tenure: Where possible new employees DO NOT operate dangerous equipment in initial employment period
Task assignment	Is there a procedure in place to stop new employees being given tasks which other employees see as undesirable and which could potentially be risky?	Task allocation is controlled by management
Workload and performance expectations	Is there a procedure in place to control the new employee's workload and performance expectations?	Policy where workload and performance expectations are relaxed during new employee's initial period of employment
Working hours and Scheduling	Is there a procedure in place to consider safety when scheduling the new employees working hours?	If possible restrict a new employee to a maximum of 8-h shift during their initial period of employment
Environmental variance	Where there is environmental variance, is there a strategy in place to ensure the new employee is not asked to work immediately in the most hazardous environment?	Assess environmental risk for all jobs
		Where possible avoid allocating new employees to perform the job in high-risks environments during their initial employment period
Co-worker behavior	Are the characteristics of the new employee's co-workers being considered when they are assigned to a work group or team?	Where possible: assign new employees to work groups that have experienced (senior) co-workers
Supervision	Is a procedure in place to ensure supervision of new employees is appropriate, and for sufficient duration to ensure their safety?	Develop a supervision model which provides specifically for the supervision needs of new employees
		Avoid informally allocating new employee supervision to co-workers
Employee silence and safety voicing	Does the organization have a safety voicing culture?	Regularly assess safety voicing
	Are new employees being provided with all the safety information they need?	Put in place a safety-specific exit voicing process

4.3 Conclusions

By identifying the normal and additional (controllable) risk factors associated with a job, an organization can minimize a new employee's exposure to controllable risk factors and reduce their chances of an accident in their initial employment period. Eight aspects of work which can contribute safety risk for a new employee have been discussed. All of the factors are controllable, but admittedly some are more easily dealt with than others. Each of the 8 aspects needs to be considered, and policy put in place to deal with each before a new employee arrives on the job. Table 4.1 shows a job risk profile checklist which could be used to consider each of the 8 risk factors. A key question or questions are proposed for each risk factor, and the far right-hand column of Table 4.1 has a possible course of action which could be taken to eliminate or minimize the risk factor. It is important to note that the 8 factors are particularly important for safety in a new employee's initial period of employment, which is roughly their first 3 months on the job.

References

Arunraj, N. S., & Maiti, J. (2007). Risk-based maintenance—Techniques and applications. *Journal of Hazardous Materials, 142*, 653–661.

Arunraj, N. S., & Maiti, J. (2010). Risk-based maintenance policy selection using AHP and goal programming. *Safety Science, 48*, 238–247.

Bell, J. L., & Grushecky, S. T. (2006). Evaluating the effectiveness of a logger safety training program. *Journal of Safety Research, 37*, 53–61.

Bertolini, M., Bevilacqua, M., Ciarapica, F. E., & Giacchetta, G. (2009). Development of risk-based inspection and maintenance procedures. *Journal of Loss Prevention in the Process Industries, 22*, 244–253.

Bjerkan, A. M. (2010). Health, environment, safety culture and climate—analyzing the relationships to occupational accidents. *Journal of Risk Research, 13*(4), 445–477.

Brinsfield, C. T. (2013). Employee silence motives: Investigation of dimensionality and development of measures. *Journal of Organizational Behavior, 34*, 671–697.

Burt, C. D. B., Cottle, C., Näswall, K., & Williams, S. (2013). Capturing safety knowledge: Using a safety-specific exit survey. Paper presented at 14th European Conference on Knowledge Management. Kaunas, Lithuania.

Burtscher, M. J., & Manser, T. (2012). Team mental models and their potential to improve teamwork and safety: A review and implications for future research in healthcare. *Safety Science, 50*, 1344–1354.

Cannon-Bowers, J. A., Salas, E., & Converse, S. (1993). Shared mental models in expert team decision making. In N. J. Castellan (Ed.), *Individual and group decision making* (pp. 221–246). NJ: Lawrence Erlbaum Associates, Hillsdale.

Christian, M. S., Bradley, J. C., Wallace, J. C., & Burke, M. J. (2009). Workplace safety: A meta-analysis of the roles of person and situation factors. *Journal of Applied Psychology, 94*(5), 1103–1127.

Clarke, S. (1998). Organizational factors affecting the incident reporting of train drivers. *Work and Stress, 12*(1), 6–16.

Clarke, S. (2006). The relationship between safety climate and safety performance: A meta-analytic review. *Journal of Occupational Health Psychology, 11*(4), 315–327.

Conchie, S. M., Taylor, P. J., & Donald, I. J. (2012). Promoting safety voice with safety-specific transformational leadership: The mediating role of two dimensions of trust. *Journal of Occupational Health Psychology, 17*(1), 105–115.

Cree, T., & Kelloway, E. K. (1997). Responses to occupational hazards: Exit and participation. *Journal of Occupational Health Psychology, 2*, 304–311.

Dembe, A., Erikson, J., Delbos, R., & Banks, S. (2005). The impact of overtime and long work hours on occupational injuries and illnesses: New evidence from the United States. *Environmental Medicine, 62*, 588–597.

Dorrian, J., Baulk, S. D., & Dawson, D. (2011). Work hours, workload, sleep and fatigue in Australian rail industry employees. *Applied Ergonomics, 42*, 202–209.

Driscoll, T. R., Ansari, G., Harrison, J. E., Frommer, M. S., & Ruck, E. A. (1995). Traumatic work-related fatalities in forestry and sawmill workers in Australia. *Journal of Safety Research, 26*(4), 221–233.

Feinberg, R. A., & Jeppeson, N. (2000). Validity of exit interviews in retailing. *Journal of Retailing and Consumer Services, 7*, 123–127.

Folkard, S., & Tucker, P. (2003). Shift work, safety and productivity. *Occupational Medicine, 53*, 95–101.

Giacalone, R. A., Knouse, S. B., & Montagliani, A. (1997). Motivation for and prevention of honest responding in exit interviews and surveys. *Journal of Psychology, 131*, 438–448.

Gillen, M. D., Baltz, D., Gassel, M., Kirsch, L., & Vaccaro, D. (2002). Perceived safety climate, job demands, and coworker support among union and nonunion injured construction workers. *Journal of Safety Research, 33*, 33–51.

Gordon, M. E. (2011). The dialectics of the exit interview: A fresh look at conversations about organizational disengagement. *Management Communication Quarterly, 25*, 59–86.

Hänecke, K., Tiedemann, S., Nachreiner, F., & Grzech-Sukalo, H. (1998). Accident risk as a function of hour at work and time of day as determined from accident data and exposure models for the German working population. *Scandinavian Journal of Work, Environment and Health, 24*, 43–48.

Hofmann, D. A., Jacobs, R., & Landy, F. (1995). High reliability process industries: Individual, micro and macro organizational influences on safety performance. *Journal of Safety Research, 26*, 131–149.

Holcroft, C. A., & Punnett, L. (2009). Work environment risk factors for injuries in wood processing. *Journal of Safety Research, 40*, 247–255.

Jaffa, N., Abdul-Tharim, A. H., Mohd-Kamar, I. F., & Lop, N. S. (2011). A literature review of ergonomics risk factors in construction industry. *Procedia Engineering, 20*, 89–97.

Khan, F. I., & Haddara, M. M. (2003). Risk-based maintenance (RBM): A quantitative approach for maintenance/inspection scheduling and planning. *Journal of Loss Prevention in the Process Industries, 16*, 561–573.

Kincaid, W. H. (1996). Safety in the high-turnover environment. *Occupational Health and Safety, 65*, 22–25.

Lenné, M. G., Salmon, P. M., Liu, C. C., & Trotter, M. (2012). A systems approach to accident causation in mining: An application of the HFACS method. *Accident Analysis and Prevention, 48*, 111–117.

Lombardi, D., Folkard, S., Willets, J., & Smith, G. (2010). Daily sleep, weekly working hours and risk of work related injury: US national health interview survey (2004–2008). *Chronobiology International, 27*(5), 10103–11030.

Marks, M. A., Zaccaro, S. J., & Mathieu, J. E. (2000). Performance implications of leader briefings and team-interaction training for team adaption to novel environments. *Journal of Applied Psychology, 85*(6), 971–986.

Mathieu, J. E., Heffner, T. S., Goodwin, G. F., Cannon-Bowers, J. A., & Salas, E. (2005). Scaling the quality of teammates' mental models: Equifinality and normative comparisons. *Journal of Organizational Behavior, 26*(1), 37–56.

McLaughlin, A. C., & Mayhorn, C. B. (2011). Avoiding harm on the farm: Human factors. *Gerontechnology, 10*, 26–37.

Milliken, F. J., Morrison, E. W., & Hewlin, P. F. (2003). An exploratory study of employee silence: Issues that employees don't communicate upwards and why. *Journal of Management Studies, 40*(6), 1453–1476.

Morrongiello, B. A., Pickett, W., Berg, R. L., Linneman, J. G., Brison, R. J., & Marlenga, B. (2008). Adult supervision and pediatric injuries in the agricultural worksite. *Accident Analysis and Prevention, 40*, 1149–1156.

Nachreiner, F. (2001). Time on task effects on safety. *Journal of Human Ergology, 30*, 97–102.

Probst, T. M., & Estrada, A. X. (2009). Accident under-reporting among employees: Testing the moderating influence of psychological safety climate and supervisor enforcement of safety practices. *Accident Analysis and Prevention, 42*(5), 1438–1444.

Reason, J. T. (1990). *Human error.* Cambridge: Cambridge University Press.

Reason, J. T. (Ed.). (1997). *Managing the risks of organisational accidents.* Aldershot: Ashgate.

Ring, J. K. (2010). The effect of perceived organizational support and safety climate on voluntary turnover in the transportation industry. *International Journal of Business Research and Management, 1*, 156–168.

Rosa, R. (1995). Extended workshifts and excessive fatigue. *Journal of Sleep Research, 4*, 51–56.

Rosness, R., Blakstad, H. C., Forseth, U., Dahle, I. B., & Wiig, S. (2012). Environmental conditions for safety work—Theoretical foundations. *Safety Science, 50*, 1967–1976.

Salminen, S. (2010). Shift work and extended working hours as risk factors for occupational injury. *The Ergonomic Open Journal, 3*, 14–18.

Slappendal, C., Laird, L., Kawachi, I., Marshell, S., & Cryer, C. (1993). Factors affecting work related injury in forestry workers: A review. *Journal of Safety Research, 24*, 19–32.

Tucker, S., Chmiel, N., Turner, N., Hershcovis, S., & Stride, C. B. (2008). Perceived organizational support for safety and employee safety voice: The mediating role of co-worker support for safety. *Journal of Occupational Health Psychology, 13*(4), 319–330.

Viscusi, W. K. (1979). Job hazards and worker quit rates: An analysis of adaptive worker behaviour. *International Economic Review, 20*, 29–58.

Webb, G. R., Redman, S., Wilkinson, C., & Sanson-Fisher, R. W. (1989). Filtering effects in reporting work injuries. *Accident Analysis and Prevention, 21*(2), 115–123.

Wiegmann, D. A., & Shappell, S. A. (2003). *A human error approach to aviation accidents analysis: The human factors analysis and classification system.* Aldershot, UK: Ashgate.

Withey, M. J., & Cooper, W. H. (1989). Predicting exit, voice, loyalty, and neglect. *Administrative Science Quarterly, 34*(4), 521–539.

Wright, C. (1986). Routine deaths: Fatal accidents in the oil industry. *Sociological Review, 4*, 265–289.

Zhaoyang, T., Jianfeng, L., Zongzhi, W., & Weifeng, H. (2011). An evaluation of maintenance strategy using risk based inspection. *Safety Science, 49*, 852–860.

Zohar, D. (1980). Safety climate in industrial organizations: Theoretical and applied implications. *Journal of Applied Psychology, 65*, 96–102.

Zohar, D., & Tenne-Gazit, O. (2008). Transformational leadership and group interaction as climate antecedents: A social network analysis. *Journal of Applied Psychology, 93*, 744–757.

Chapter 5
The Influences of Recruitment Processes and Selection Predictors on New Employee Safety

5.1 Introduction

This chapter discusses the section of Fig. 1.1 labeled *recruitment and selection processes*, these being the processes which an organization uses to hire a new employee into a job (e.g., job recruitment advertisement, application blank, applicant interview, ability testing, personality testing). For jobs where safety is an issue, job applicant recruitment and assessment should include a consideration of the job's safety risks, and the applicant's safety attitudes, and safety-related competencies. Research has clearly shown that attitudes toward safety are associated with safety behavior. An individual's safety attitudes reflect their beliefs regarding safety policies, procedures, and practices (Neal and Griffin 2004; Rundmo and Hale 2003), and safety attitudes have been found to have a direct effect on risk taking (Rundmo 1996), and on safety compliance behaviors (McGovern et al. 2000). Furthermore, in order to work safely employees need to have the competencies (knowledge, skills, and abilities) which the job's tasks require.

Organizations vary in the amount of focus they place on predicting job applicant's safety behavior, and/or their ability to work safely. Furthermore, the ability of measures to predict safety attitudes and safety-related competencies varies. That is, while measures may be used during selection to predict safety, they may in fact have very limited predictive power. Complicating matters further are the perceptions of employees of their organization's recruitment and selection processes. Unfortunately, it is likely that employees exposed to a selection predictor (or who know that such a predictor is used to select new employees) will simply assume it is a validity and reliable measure, and as such is making a positive difference. In the case of selecting individuals into a high-safety-risk job, using predictors which focus on safety, the positive difference is assumed to be a new employee who is more likely to work, or be able to work, safely. The same may be said for the organization using the predictor. They assume it is actually predicting accurately.

© Springer International Publishing Switzerland 2015
C.D.B. Burt, *New Employee Safety*,
DOI 10.1007/978-3-319-18684-9_5

Thus, employees within an organization form opinions about their organizations recruitment and selection processes. These opinions will be partly based on their own experiences when they were recruited into the organization and partly based on what they 'see' and 'hear' about how other people were recruited and selected. Over several studies (e.g., Burt and Hislop 2013; Burt et al. 2009; Burt and Stevenson 2009), we have shown several worrying relationships between employees' perceptions of organizational processes and their perceptions of new employees. A brief summary of the key findings in these studies is provided in Table 5.1. The research suggests employees may place a considerable degree of trust in the ability of organizational processes to achieve positive safety outcomes. That is, to deliver new employees, that will work safely.

Figure 5.1 shows the probable causal nature of the research findings presented in Table 5.1. The top box in Fig. 5.1 represents employee's perceptions of their organization's recruitment and selection processes. Positive perceptions of these processes were found in all three studies to be associated with an increase in trust in new employees to work safely, a decrease in the perceived safety risk associated with new employees, and a reduction in perceived safety risk from new employees was associated with a reduction in behaviors toward the new employee to ensure their (and everyone's) safety.

The associations shown in Fig. 5.1 are nicely explained by Wilde's risk homoeostasis theory (RHT) (see Glendon et al. 1996; Wilde et al. 2002; Simonet and Wilde 1997). RHT predicts that as safety features are added to a system, individuals will increase their risk taking. A number of studies have shown how

Table 5.1 Correlations between trust in selection processes and, trust in new employees, perceived risk from new employees, and employees compensatory behaviors toward new employees obtained in 3 studies

Study	Sample	Correlation between trust in selection processes and trust in new employees to work safely	Correlation between trust in new employees and perceived safety risk from new employees	Correlation between perceived safety risk from new employees and compensatory behaviors to ensure new employee safety
Burt et al. (2009)	128 forestry workers	0.23^{**}	-0.20^{*}	0.33^{**}
Burt and Stevenson (2009)	154 professional firefighters	0.29^{**}	-0.24^{*}	0.43^{**}
Burt and Hislop (2013)	118 employees in high-risk jobs from 5 organizations	0.29^{**}	-0.13	0.43^{**}

$^{*}P < 0.05$; $^{**}P < 0.01$

Fig. 5.1 Associations
between organizational
processes, and employees'
perceptions of, and reactions
to, new employees

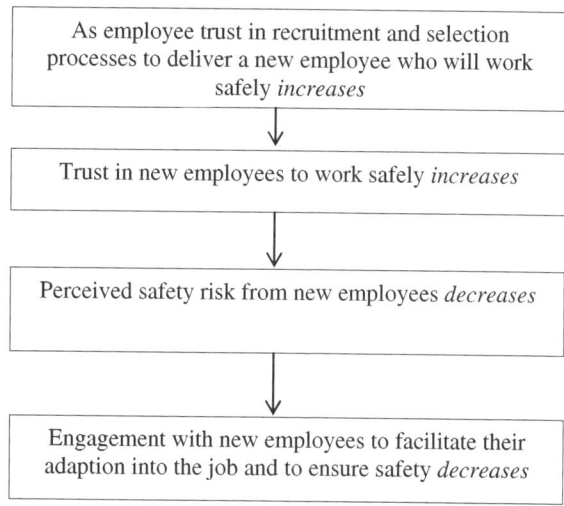

RHT operates. Studies by Aschenbrenner and Biehl (1994) and Grant and Smiley (1993), both of which measured driver behavior in vehicles with and without antilock brakes, clearly showed that drivers changed their behavior (it became more risky) in response to the presence of the antilock brake system. In other words, the drivers increased their risk taking (known as behavioral adaption) in the presence of a system which they assumed would provide a degree of safety (protection). In the case of the relationships shown in Fig. 5.1, employees seem to be adapting their behavior toward new employees (taking more risk) based on their assumption that the organization's recruitment and selection processes are providing a degree of safety (protection).

It is also not too difficult to understand why employees might assume that an organizations recruitment and selection processes are having positive outcomes. First, most employees are *unlikely* to have an in-depth understanding of the limitations of measures used to predict behavior. Secondly, safety is part of the psychological contract that forms between an employee and their organization (Walker 2013; Walker and Hutton 2006). One of the basic relationships in the psychological contract is that the employee and the employer have reciprocal safety obligations. It might be expected that employees expect their employer will do everything they can to ensure they are not exposed to safety risks from a new employee. Therefore, the psychological foundation for employees trusting an organization's recruitment and selection processes may reside in their psychological contract. Although there are clear safety benefits from trust, unearned trust, which is essentially what trust based on organizational process delivers, can have a negative impact on safety. Such trust can reduce an employees' inclination to monitor and safeguard and can decrease their judgment of new employees based on *their* behaviors (Conchie and Donald 2008; McEvily et al. 2003). Overall, inaccurate perceptions of the effectiveness of an organization's recruitment and selection processes may contribute to the new employee safety risk. Trust may stop a co-worker keeping an eye on a new employee—or even prompt a worker to leave a new employee in a situation where they could well get into difficulties—particularly if they attempt to do something for which they are not trained or experienced.

Of course, if recruitment and selection activities are achieving their objectives, are accurately predicting new employee's safety attitudes and ability to work safely, employee's perceptions that these processes do help ensure safety will be correct. Furthermore, research suggests that sometimes this is a correct assumption to hold. For example, studies by Cohen (1977) and Smith et al. (1978) found that organizations with low accident rates also had more elaborate selection systems. However, achieving accurate prediction in a selection system is not an easy task. Furthermore, even if an organization's recruitment and selection system is having a positive influence on new employee safety, it is not ideal for employees to be less careful around new employees. Chapter 7 discusses in detail processes which can be adopted by an organization to ensure that employees engage in an appropriate way with new employees, and which encourage employees to respect the safety risks that new employee bring into the work place.

There is, however, a balancing act required here. While it is not good for employees to overestimate the positive impact of recruitment and selection processes on new employee safety, positively regarding these organizational processes may be associated with an increase in perceptions of management's commitment to safety. While there appears to be no research which has directly addressed this issue, it is possible, even likely that employees perceptions of recruitment and selection processes are part of the basis upon which their perception of management's commitment to safety is based. If this is the case, it has very important implications as numerous studies have identified the importance of perceptions of management's commitment to safety. For example, Cui et al. (2013) found that safety behaviors where mediated by perceptions of management's commitment to safety.

The following sections examine two processes associated with recruitment activities (i.e., job analysis, and realistic safety preview), and a range of selection predictors (application blank, applicant interview, cognitive, physical, psychomotor, sensory/perceptual ability testing, personality testing, and attitude measurement) which an organization can use to help predict job applicant (new employee's) safety behavior, and overall their ability to work safely. Where appropriate, recommendations on how recruitment and selection processes can be used to improve new employee safety are discussed. Finally, this chapter examines how employees' perceptions of organizational processes can be made more realistic.

5.2 Defining What to Communicate and What to Measure

A critical question in the development of a recruitment and selection process is what information should be delivered to job applicants at the time of recruitment, and what should be measured in job applicants at the time of selection. Ramsey's *accident sequence model* (see Ramsey 1985) nicely articulates how important human abilities can be for safety. Ramsey notes how a hazard must be perceived (which requires sensory and perceptual abilities), how a perceived hazard must be cognized (or understood to be a hazard) which involves mental abilities, how a decision must be made to avoid the hazard which can be influenced by attitudes and personality, and finally how the individual must have the anthropometric characteristics and motor skills required to avoid the hazard. A breakdown in any aspect of the sequence, and there is a chance that exposure to a hazard will result in an accident. Ramsey's model provides a solid foundation for an employee selection program to include measures of the human abilities required to ensure an employee's safe performance of their job. This conclusion is supported by the study conducted by Ford and Wiggins (2012) where cognitive ability and skill mediated the relationship between physical hazard and injury/incident rate. Furthermore, it highlights the importance of communicating information on the abilities required to perform a job safely during the recruitment phase. While Ramsey's model provides a clear general argument, the use of job analysis is required to identify the specific knowledge, skills, and abilities required for a job.

5.3 Job Analysis

All recruitment and selection activities should begin with a job analysis. Job analysis allows the task requirements of a job to be precisely determined. Furthermore, job analysis allows the safety risks associated with a job to be determined, and also the identification of the knowledge, skills, and abilities required to work, both safely, and at a satisfactory performance level. It is well established that the occupation or job a person is performing substantially influences accident vulnerability (Ford and Wiggins 2012). In other words, it is vital for safety, for the specific hazards and risks associated with a job to be identified, and conducting a job analysis is an approach which can be used to collect this information. Without the essential information which job analysis provides, it is impossible to provide job applicants with a realistic safety preview for the job (see Chap. 3, Sect. 3.7.2.1), and difficult to know what competencies a new employee needs to bring to the job, and therefore what should be measured in a selection program.

There are a number of very good books which deal with job analysis methods (e.g., Brannick et al. 2007; Prien et al. 2009; Wilson 2012), and also very good guidelines on how to avoid error in job analysis data (e.g., Morgeson and Campion 1997). In general, the aim of job analysis is to produce two documents, a job description and a person specification. The job description document contains all of the information relating to a job's roles and tasks, a description of the context within which these are performed, and performance expectations, and benefits. Where appropriate, the job description should also include a specific section on the job safety risks and hazards (see Fig. 3.1 in Chap. 3). The other main document that is produced by job analysis is the person specification. This document describes the essential knowledge, skills, abilities, and other characteristics which are required to perform the job.

In addition to the literature on job analysis, there is also a body of work which has focused solely on methods for identifying the hazards and risks associated with a job. As some examples, techniques such as job safety analysis, also referred to as job hazard analysis (Chao and Henshaw 2002), construction job safety analysis (Rozenfeld et al. 2010), and construction hazard assessment with spatial and temporal exposure (Rosenfeld et al. 2009) have been extensively discussed. These techniques are very useful for safety management, and among other things can be used to generate safety-related information which can be used to determine the essential knowledge, skills, abilities, and other characteristics which are required to perform the job safely.

Job description and person specification documents that contain safety information can clearly provide the foundation for the development of a recruitment and selection program which has at least the possibility of a successful outcome. Further, Thompson and Thompson (1982) provide an excellent review of the steps required to help ensure that courts accept job analysis information as the foundation of selection predictor development and or selection decisions. Furthermore, a job description that includes a section on safety can be used in the expectation setting processes as discussed in Chap. 3, Sect. 3.7.2. In contrast, if there has been no systematic attempt to understand what the requirements are to perform a job in a safe manner, it is unlikely that the recruitment and selection system will be delivering the safety benefits that it potentially could. Furthermore, it is likely that employees' trust in these processes to deliver a safe new employee may be somewhat misplaced.

5.4 Recruitment Processes

Broadly speaking recruitment processes are those activities which an organization uses to obtain a pool of applicants for a vacant position (e.g., advertising a vacancy, providing interested individuals with information about the job via a job description and person specification). A number of models of the recruitment process have been developed, with the model developed by Breaugh et al. (i.e., Breaugh 2008, 2013; Breaugh and Starke 2000) representing 'state-of-the-art' practice. A key feature of Breaugh's model is the integration of a realistic job preview into the recruitment process. A realistic job preview provides job applicants with information about job tasks, job environment, and the competencies required to perform the job. Numerous studies have examined the realistic job preview process, although very little if any research appears to have examined it specifically from a safety perspective. From a safety perspective, a realistic job preview would involve clearly explaining to job applicants via the recruitment advertising, job description and person specification documents, and in face to face discussions, all of the safety issues associated with a job. This would include the risks and hazards associated with the job and environment in which it is performed, and the safety behavior that the individual would need to engage with in order to maintain safety.

In general, applicants exposed to a realistic job preview react in a number of ways. Some applicants decide not to pursue the vacant job deciding the job is not for them or that they do not have the required competencies to perform the job. Those that continue with the application process and are hired are likely to have a higher level of role clarity, more trust in the organization, more commitment to the organization and job, more job satisfaction, and are likely to stay in the job longer. Furthermore, new employees recruited into a job which has safety issues which have been clearly explained via a realistic job preview process are likely to be better prepared for the job's safety demands. This is achieved via the realistic job preview information providing clarity as to exactly what tasks are involved, exactly what the associated safety factors are, and exactly what is expected of the employee in terms of safety. Figure 5.2 is a representation of a realistic safety preview and its expected effects on job applicants and new employees.

Using a recruitment process that includes a realistic safety preview is likely to greatly increase the chances that an organization will employ a new employee that is able to work in a safe manner. At the very least, the process should help remove applicants who decide themselves that they do not have the competencies required

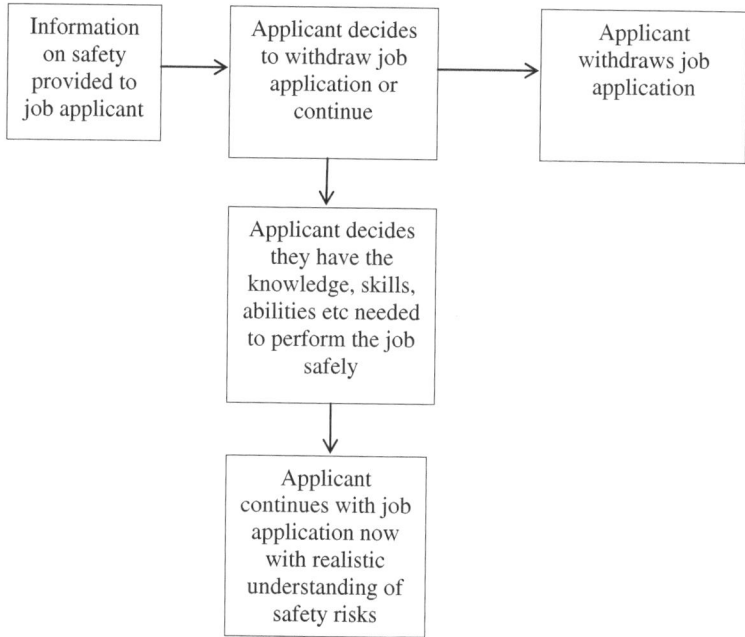

Fig. 5.2 Decisions associated with *realistic safety preview* use at the time of new employee recruitment

to do the job in a safe manner. This is no minor outcome, as in many respects it is the individual that has the most accurate understanding of their abilities. To be sure, there will be other consideration such as economic needs, which may prompt individuals' to purse jobs for which they lack essential competencies, but hopefully other steps in the selection process will identify these individuals. Unfortunately, it is unclear whether organizations are using realistic safety previews (see Chap. 3 sect. 3.7.2.1 for related research). Certainly it is not a topic which has been given much research attention. If there has been no systematic attempt to integrate a realistic safety preview into a recruitment process, it is unlikely that the recruitment system will be delivering the potential safety benefits that it could be. Furthermore, it is likely that employees trust in recruitment processes may be somewhat misplaced.

5.5 New Employee Selection Predictors

Having obtained a group of individuals who have applied for a job via the recruitment processes, the next step is to determine which individual or individuals from among the applicant pool would be best to employee. At this stage, the organization can assess the job applicants using a number of different selection predictors. The key component of any selection predictor is its criterion-related validity. Basically, criterion-related validity is the degree to which the predictor (measure) is able to predict variance in job (or safety) performance. For example, if a job interview contains questions about an applicant's safety compliance and participation, how well do the applicant's answers to the questions reflect the individual's on-the-job safety participation and compliance? From a research perspective, it is possible to answer this question empirically by conducting a validation study where scores for the interview question answers are correlated with some post-employment measure of the new employee's safety participation and compliance (perhaps given by a supervisor or co-worker). Unfortunately, validation work is rarely performed. Below I examine various types of predictor which may be used as part of a selection program to predict new employee safety, and in particular look for evidence of criterion-related validity. If the predictors used in a selection program have no criterion-related validity, the selection system will not be delivering the potential safety benefits that it could be. Furthermore, it is likely that employees' trust in the selection processes to deliver a safe new employee will be misplaced.

5.5.1 Application Blank

Application blanks are useful when a large number of individuals apply for a relatively small number of jobs. However, their usefulness is very dependent on the processes used to design the applicant blank and the associated scoring process. An application blank can measure essential knowledge and skills, can be objectively scored, and can be used to quickly screen applicants in and out of a selection process. Furthermore, the use of an application blank is an ideal way to measure job applicant experience. Chapter 3 discusses in detail the nature of experience and its relationship with being classified as experienced. Questions which can be used to assess a job applicants' experience are also provided in Chap. 3 (see Table 3.3). These questions can easily be incorporated into an application blank or used in an interview.

While many organizations still request job applicants' to provide a CV, the use of an application blank has many advantages over the traditional CV. Wood and Payne (1998) offer a competency-based approach to application blank design. Using a competency-based format, the application blank only requests information on job-related competencies, as opposed to the CV which may well contain a lot of information which is not relevant to the selection process. Further, an applicant blank can be structured such that the required information is presented by the applicant in specific sections, whereas the unstandardized nature of CVs can require a recruiter to undertake a lot of searching in order to identify the required information.

As with all measures used to select individuals for employment, it is vital to ensure that the application blank does not include questions which are not relevant for, or are not related to, job performance. Kethley and Terpstra (2005) examined court cases involving the alleged illegal use of questions in application blanks. Their research highlights many questions which the courts have found to be illegal or discriminatory. Their research further highlights the value of performing a job analysis as the first step in determining the actual competencies which are required to perform a job. It seems clear that a well-designed, and objectively scored, application blank may prove to be a valid and reliable predictor of a job applicant's job performance potential. Equally, aspects of an application blank, and perhaps specifically, a highly structured and detailed section on experience, could be a valid and reliable predictor of a job applicant's potential to perform a job in a safe way, or to adapt more quickly to a job's hazards and risks.

If used, it is likely that an application blank will form a hurdle in the selection process. That is some applicants will be rejected based on their application blank information, while others will pass the application blank stage and continue on in the selection process. As noted employees in the organization will make assumptions about the application blank, and about its ability to make valid and reliable *continue* or *reject* decisions. Employees are likely to assume that the decisions that are made are valid and reliable and that those that pass the application blank hurdle are indeed likely to be employees that can perform in the job, and do this in a safe manner. Given that a number of other very important perceptions and behaviors flow from these assumptions (as shown in Fig. 5.1), it is very important that the application blank and its scoring actually does provide a valid and reliable assessment of job applicants' ability to perform the job in a safe way.

5.5.2 Employment Interviews

Use of an interview to select employees is common place. As such hundreds of studies have been conducted on various aspects of the employment interview process. It is now well established that using a structured employment interview process improves criterion-related validity (Schmidt and Hunter 1998). Furthermore, a number of very good papers have been published on the process of developing a structured employment interview (e.g., Barclay 2001; Campion et al. 1997; Huffcutt 2011; Levashina et al. 2014). The research evidence is clear that developing and using a highly structured interview process can help provide a valid and reliable measure of a job applicant's ability to perform a job. However, like all selection measures, an interview is more suited to the measurement of some competencies than others.

While an interview can measure a range of safety-related dimensions (see Chap. 3, Tables 3.3 and 3.4 for questions that can be used to measure an applicant's experience and expectations, respectively, in an employment interview), they are not particularly well suited to the measurement of aspects of safety such as safety motivation, participation, and compliance. The key problem is the possibility that the applicant will respond in a socially desirable way, and their response may have little relationship to their future on-the-job behavior. Rather than questioning job applicants directly about their safety attitudes (e.g., asking questions like 'Should safety have a high priority?'), a structured interview can make use of either situational or behavioral questions to gain a perspective on the applicant's safety attitudes. Both situational and behavioral questions are based on scenarios or critical incidents which come directly from the job which the individual is being recruited for. The primary difference is that when using the behavioral format, the individual is asked what they have done in the described situation in the past, while in the situational format the applicant is asked what they would do in the particular situation.

Flanagan (1954) developed the procedure known as the critical incident technique. This relatively simple process involves interviewing job incumbents and asking for descriptions of critical incidents in their job, and also asking what they did in the particular situation. Critical incident information could also be obtained by supervisors keeping a record of situations they have observed, and employees' responses to the situation. Thus, a critical incident represents a specific job situation and a particularly effective response to that situation. The critical incident technique can easily be applied to gather safety-specific examples. A sample of employees would be asked to describe a situation which had a safety aspect and then to describe how the safety issue was handled or resolved. Of course it is necessary to ensure that the response to the situation is indeed the correct response in that it is what the organization would want an employee to do when the particular safety situation occurred. Once a number of these critical incidents have been identified, they can be formed into employment interview questions. The job applicant is presented with the question (or scenario) and is assessed on their description of how they would (or have) handle or responded to the situation, and in particular how closely their answer matches what the organization expects employees to do under these circumstances.

A structured employment interview could also be used to measure knowledge which is required to work in a safe manner. Obviously this knowledge will be idiosyncratic to the job in question. The job-relevant knowledge which an applicant has is often inferred based on their work and education history. While there should, for example, be a relationship between education and what an applicant's knows, it is not always wise to make this assumption. Furthermore, employees are going to expect (assume) that new employees do have the knowledge necessary to perform their job in safe manner.

5.5.3 Cognitive Ability Testing

Jobs vary in terms of the cognitive abilities required for effective and safe performance. It is well established that general cognitive ability is a valid and reliable predictor of overall job performance, job-knowledge acquisition, and training performance (Schmidt and Hunter 1998). Furthermore, evidence of positive relationships between cognitive ability and safety behavior (or the ability to work safely) is mounting (e.g., Ford and Wiggins 2012; Postlethwaite et al. 2009). While the measurement of general mental ability is undoubtedly useful for the prediction of employee behavior, a more fine-grained analysis of cognitive abilities is possible. Fleishman and Reilly (1995) define 21 different cognitive abilities: oral comprehension, written comprehension, oral expression, written expression, fluency of ideas, originality, memorization, problem sensitivity, mathematical reasoning, number facility, deductive reasoning, inductive reasoning, information ordering, category flexibility, speed of closure, flexibility of closure, spatial orientation, visualization, perceptual speed, selective attention, and time sharing. New employee safety (and workplace safety in general) is dependent on identifying the specific cognitive abilities which are required for safe performance of the job, measuring these abilities in job applicants, and selecting applicants that have a suitable ability level. For example, McMullan and Lea (2010) identified numerical ability as a requirement for nurses as it provides for the accurate and safe administration of medications (also see Grandell-Niemi et al. 2003). Without a suitable level of numerical ability, nurses can put patients at risk through inappropriate medication administration.

There are a number of issues which can have a negative impact on the accuracy of cognitive ability testing. Any issue which has a negative impact on the accuracy of the assessment will also have a negative impact on the accuracy of the decisions made based on the assessment information. While there are numerous textbooks devoted to psychological testing, the main issues which can influence assessment accuracy are relate to appropriate test selection, test validity and reliability, test administration, and test score interpretation. It is beyond the scope of this book to go into these issues in detail. However, the key point is that employees in an organization are likely to assume that a test used to assess abilities in job applicants, and ultimately to make a hiring decision, is in fact a valid and reliable tool, and as such, its use is helping to ensure that new employees will be able to work safely.

5.5.4 Anthropometric Characteristics, and Physical, Psychomotor, Sensory/Perceptual Ability

Anthropometric science is the study of variance in human body dimensions, and how this needs to be incorporated into design solutions to provide for efficient and safe operation of equipment and systems (Hsiao 2013; Lee and Bro 2008). Understanding variation in human body dimensions and strength provides for the correct design of equipment. Given the variation which can be identified in various body dimensions, it is normal practice for equipment to be designed to accommodate 95 % of a population. Thus, there will be occasions where a specific design feature will not suit a specific individual. For example, the cab of a truck may not suit an individual that is extremely tall. Arguably many jobs which have an associated safety consideration will require a consideration of applicants' anthropometric characteristics (Grandjean 1985).

Fleishman and Reilly (1995) define nine physical abilities (i.e., static strength, explosive strength, dynamic strength, trunk strength, extent flexibility, dynamic flexibility, gross body coordination, gross body equilibrium, and stamina), nine psychomotor abilities (i.e., control precision, multilimb coordination, response orientation, rate control, reaction time, arm-hand steadiness, manual dexterity, wrist-finger speed, and speed of limb movement), and eleven sensory/perceptual abilities (i.e., near vision, far vision, visual color discrimination, night vision, peripheral vision, depth perception, glare sensitivity, auditory attention, sound localization, speech recognition, and speech clarity). As with anthropometric characteristics, a job will vary in terms of its physical, psychomotor, and sensory/perceptual ability requirements. Arguably many jobs which have safety considerations have a high level of physical, psychomotor, and perceptual demand. Assessment of the physical, psychomotor, and perceptual demands has advantages; in that, well-developed tests are normally available and are typically objectively scored. Furthermore, physical, psychomotor, and perceptual ability tests can be developed for specific jobs (see Pelot et al. 1999 for a forcible entry test for firefighters) providing even better measurement outcomes.

5.5.5 Personality Testing

Personality testing of job applicants has received considerable research attention. A number of review papers (e.g., Arthur et al. 2001; Morgeson et al. 2007; Scroggins et al. 2009) and meta-analysis studies (e.g., Barrick and Mount 1991) have shown varying degrees of support for the relationship between personality dimensions and job performance. In addition to the focus on the relationship between personality and job performance, within the field of safety there has been a long history of researchers suggesting that an individual's personality may be related to their involvement in accidents. At the extreme end of this discussion is the concept of *accident proneness*.

A significant amount of research attention has been given to the issue of accident proneness. The basic premise, popular in the early part of last century (see Green and Woods 1919) was that human error (in at least some accidents) may be attributable to a constellation of personality or individual difference variables which predispose some individuals (the accident prone) to be more likely to cause or be involved in an accident. While the idea of accident proneness lost favor in the mid-part of the twentieth century, it still receives sporadic research attention (e.g., Dahlback 1991; Visser et al. 2007). Perhaps the appealing aspect to the topic of accident proneness is the idea that the propensity to engage in behaviors which can result in an accident is sometimes related to risk taking, and some authors consider risk taking is a personality trait (e.g., Dahlback 1990; Eysenck and Eysenck 1977). The meta-analysis conducted by Visser et al. (2007) provides a useful overview of 79 studies which have empirically examined the general idea of accident proneness. Interestingly, Visser et al. (2007) concluded there was some evidence to support the existence of accident prone individuals, with their analysis showing accidents clustering within individuals at a rate higher than expected by chance. While the Visser et al. (2007) meta-analysis did not address the underlying personality characteristics of accident proneness, it does argue for the value of undertaking further investigations.

Personality has many different dimensions, and this complexity creates many difficulties for attempts to clearly define how personality is related to work performance and safety outcomes. The emergence of the Big Five personality model (McCrae and Costa 1990) has helped reduce this complexity, and the Big Five personality taxonomy has provided a useful framework for the exploration of the criterion-related validity of personality measures (e.g., Barrick and Mount 1991; Salgado 1997). Following this trend, Clarke and Robertson (2005) used meta-analysis to examine the criterion-related validity of the Big Five personality factors as predictors of accident involvement. Forty-seven studies were identified that meet the selection criteria for inclusion in the meta-analysis. Studies that had examined accidents at work and non-occupational accidents (mainly traffic accidents) were included. Overall two personality dimensions, low conscientiousness, and low agreeableness were reported to be valid and generalizable predictors of accidents in the samples examined. These results strongly suggest that job applicants with low scores on conscientiousness and agreeableness may be more likely to have an accident. A slightly different picture emerged when only occupational accidents were considered with low agreeableness and neuroticism being the best predictors of occupational accidents. A similar meta-analysis conducted by Clarke and Robertson (2008) identified low agreeableness as a valid and generalizable predictor of accident involvement. Overall, there is some reasonably consistent evidence showing that individuals identified as low on agreeableness may have more accidents.

Another line of research on personality has attempted to identify which, if any personality dimensions might be associated with individuals working safely. That is associated with safety behaviors. A key finding from this research is that conscientiousness is associated with safety motivation (Christian et al. 2009). Christian et al. (2009) identified this link in their meta-analysis examining the role of person and situation factors in safety performance. Their results also support a positive association between safety motivation and safety knowledge, and a positive relationship between both safety motivation and knowledge, and safety performance. Safety performance was measured by a combination of safety compliance (e.g., following procedures, using safety equipment) and safety participation (e.g., voicing safety issues, initiating safety changes). Thus, their data would argue that the personality dimension *conscientiousness* is an important consideration if personality is being used to select employees into jobs where safety is an issue. Hogan and Foster (2013) suggest that individuals that score low on a valid measure of conscientiousness are more likely to be inattentive, ignore the rules, and take risks, and as such are less likely to be safe employees.

It may also be the case that personality dimensions other than the Big Five may be associated with safety behavior. A number of recent studies have begun to explore this idea in some detail. For example, Probst et al. (2013) examined the ability of a personality dimension-labeled *consideration of future safety consequences* to predict employee safety. Similarly, Hogan and Foster (2013) examined the relationships between 6 personality dimensions (labeled *compliant*, *confident*, *emotionally stable*, *vigilant*, *cautious*, and *trainable*) and safety-related outcomes. While this work is important, for now the weight of evidence required to reach conclusions about these personality dimensions is not there.

Taken together the research evidence suggests an organization may be able to predict some aspects of a new employee's safety using a personality measure. In particular, the identification of job applicants that score high on both agreeableness and conscientiousness could help ensure new employee safety. It is also worth noting that personality has to be accompanied by relevant cognitive abilities for an individual to truly behave safely (e.g., Postlethwaite et al. 2009; Wallace and Vodanovich 2003). However, like other selection predictors employees may place more trust in the ability of personality measure to predict the safety behavior of new employees than is warranted. While an individual's agreeableness and conscientiousness may be associated with their safety, it would be a mistake to assume these two dimensions will guarantee safe behavior.

5.5.6 Safety Attitudes

It is very clear that an individual's attitude toward safety is likely to predict their risk-taking and safety compliance behavior. Here, I am referring to an individual's general attitude toward safety (e.g., agreement with the statement *Safety should have a high priority*), rather than safety climate attitudes which are the result of an individual's perception of their organization's policies, procedures, and practices concerning safety (e.g., agreement with the statement *Safety has a high priority at* …). Of course, an individual's general attitude toward behaving safely is likely to influence their safety climate perceptions. An individual that has a very positive attitude toward safety in general is not inclined to take risks and is generally compliant with rules, etc., which ensures their safety. Thus, if an individual's general attitude toward safety can be validly and reliably measured during the selection process, it may provide very useful information to predict on-the-job safety behavior. Unfortunately, it is not easy to measure attitudes. The difficulty is related to the ease with which socially desirable responses can be made to attitude assessing statements.

Unfortunately, it is very difficult to avoid bias responding to attitude statements. Knowledge of this should serve as a warning to organizations when they are offered measures which are claimed to measure job applicants' safety attitudes. While the statements listed in such measures may well appear to ask the right questions, the validity of the responses they generate has to be questioned. Using such measures is also likely to be particularly distorting of employee perceptions of organizational selection processes. The questions in commercial measures of job applicants' safety attitudes may look right, but there is often little or no evidence to support claims that they are valid and reliable measures. Of course employees may just simply assume they work. A very dangerous assumption indeed.

5.5.7 Commercial Products

As a final note on tools which might be used to select new employees, there are number of products on the market which claim to predict safety behaviors and safety-related outcomes. Providers of these assessment tools vary greatly in the claims that are made about their tools ability to predict employee safety behavior, and the degree of research based evidence which they provide to support these claims. Organizations using these products need to examine very carefully the nature of the instrument/measure, and the evidence that it is a valid predictor. As with other selection assessments, employees are likely to assume such measures will operate in a valid and reliable way.

5.6 Communicating a Realistic Outcome View

A step which organizations might take is to communicate accurate information about the likely outcomes of the safety-specific characteristics of their recruitment and selection processes. Such information needs to be presented in a form which is readily understood by the work force. For example, information on the criterion-related validity of selection predictors will be too technical for employees to understand. An organization, rather than attempting to build employee trust in recruitment and selection processes, can gain significant safety advantages from encouraging workers not to trust these processes. For example, an organization might inform members of a work team that '*all predictable steps have been taken to recruit new team members that will work safely. However, it is important for employees to realize that the prediction of a new employee's ability, and desire, to work safely is a difficult task, and there will be variability in the safety behavior of new employees. As such, it is not recommended that employees immediately trust a new employee's safety-related abilities and attitudes. In order to safeguard both yourself, and new employees, employees should actively monitor new employees.*'

5.7 Conclusions

The preceding sections have outlined a number of recruitment processes which can potentially facilitate new employee safety, as well as a range of selection predictors which may allow for the prediction of new employee safety behavior. Thus, it is possible to gain some safety advantages from the use of appropriate recruitment and selection processes. However, this advantage requires careful consideration of safety at the time a recruitment, and selection program is designed. It is also very important for organizations to realize that employees tend to assume that recruitment and selection processes are producing positive outcomes.

References

Arthur, W, Jr, Woehr, D. J., & Graziano, W. G. (2001). Personality testing in employment settings: Problems and issues in the application of typical selection practices. *Personnel Review, 30*(6), 657–676.

Aschenbrenner, M., & Biehl, B. (1994). Improving safety through improved technical measures? Empirical studies regarding risk compensation processes in relation to anti-lock brake systems. In R. M. Trimpop & G. J. S. Wilde (Eds.), *Changes in accident prevention: The issue of risk compensation*. Groningen, The Netherlands: Styx Publications.

Barclay, J. M. (2001). Improving selection interviews with structure: Organisations' use of behavioral interviews. *Personnel Review, 30*(1–2), 81–101.

Barrick, M. R., & Mount, M. K. (1991). The Big Five personality dimensions and job performance: A meta-analysis. *Personnel Psychology, 44*, 1–26.

Brannick, M. T., Levine, E. L., & Morgeson, F. P. (2007). *Job and work analysis: Methods, research and applications for human resources management*. Thousand Oaks: Sage Publications.

Breaugh, J. A. (2008). Employee recruitment: Current knowledge and important areas for future research. *Human Resource Management Review, 18*, 103–118.

Breaugh, J. A. (2013). Employee recruitment. *Annual Review of Psychology, 64*, 389–416.

Breaugh, J. A., & Starke, M. (2000). Research on employee recruitment: So many studies, so many remaining questions. *Journal of Management, 26*(3), 405–434.

Burt, C. D. B., Chmiel, N., & Hayes, P. (2009). Implications of turnover for safety attitudes and behaviour in work teams. *Safety Science, 47*, 1002–1006.

Burt, C. D. B., & Hislop, H. (2013). Developing safety specific trust in new recruits: The dilemma and a possible solution. *Journal of Health, Safety and Environment, 29*(3), 161–173.

Burt, C. D. B., & Stevenson, R. J. (2009). The relationship between recruitment processes, familiarity, trust, perceived risk and safety. *Journal of Safety Research, 40*, 365–369.

Campion, M. A., Palmer, D. K., & Campion, J. E. (1997). A review of structure in the selection interview. *Personnel Psychology, 50*, 655–702.

Chao, E. L. & Henshaw, J. L. (2002). *Job hazard analysis*. OSHA publication 3071 2002 (revised). Occupational safety and health administration, US Department of Labor, Washington.

Christian, M. S., Bradley, J. C., Wallace, J. C., & Burke, M. J. (2009). Workplace safety: A meta-analysis of the roles of person and situation factors. *Journal of Applied Psychology, 94*(5), 1103–1127.

Clarke, S., & Robertson, I. T. (2005). A meta-analytic review of the Big Five personality factors and accident involvement in occupational and non-occupational settings. *Journal of Occupational and Organizational Psychology, 78*, 355–376.

Clarke, S., & Robertson, I. T. (2008). An examination of the role of personality in work accidents using meta-analysis. *Applied Psychology: An International Review, 57*(1), 94–108.

Cohen, A. (1977). Factors in successful occupational safety programs. *Journal of Safety Research, 9*, 168–178.

Conchie, S. M., & Donald, I. J. (2008). The functions and development of safety-specific trust and distrust. *Safety Science, 46*, 92–103.

Cui, L., Fan, D., Fu, G., & Zhu, C. J. (2013). An integrative model of organizational safety behavior. *Journal of Safety Research, 45*, 37–46.

Dahlback, O. (1990). Personality and risk-taking. *Personality and Individual Differences, 11*, 1235–1242.

Dahlback, O. (1991). Accident-proneness and risk-taking. *Personality and Individual Differences, 12*, 79–85.

Eysenck, S. B. G., & Eysenck, H. J. (1977). The place of impulsiveness in a dimensions system of personality description. *British Journal of Social and Clinical Psychology, 16*, 57–68.

Flanagan, (1954). The critical incident technique. *Psychological Bulletin, 51*, 327–358.

Fleishman, E. A., & Reilly, M. E. (1995). *Handbook of human abilities: Definitions, measurement, and job task requirements*. Management Research Institute.

Ford, M. T., & Wiggins, B. K. (2012). Occupational-level interactions between physical hazards and cognitive ability and skill requirements in predicting incidence rates. *Journal of Occupational Health Psychology, 17*(3), 268–278.

Glendon, A. I., Hoyes, T. W., Haigney, D. E., & Taylor, R. G. (1996). A review of risk homeostasis theory in simulated environments. *Safety Science, 22*, 15–25.

Grandell-Niemi, H., Hupli, M., Leino-Kipli, H., & Puukka, P. (2003). Medication calculation skills of nurses in Finland. *Journal of Clinical Nursing, 12*, 519–528.

Grandjean, E. (1985). *Fitting the task to the man: An ergonomic approach*. London: Taylor & Francis.

Grant, B. A. & Smiley, A. (1993). Driver response to antilock brakes: A demonstration of behavioral adaption. In *Proceedings, Canadian Multidisciplinary Road safety Conference VIII, Saskatoon, Saskatchewan, June 14–16*, Saskatoon, Saskatchewan, Canada: University of Saskatchewan.

Greenwood, M., & Woods, H. M. (1919). A report on the incidence of industrial accidents with special reference to multiple accidents. *Report 4, Industrial Fatigue Research Board.*

Hogan, J., & Foster, J. (2013). Multifaceted personality predictors of workplace safety performance: More than conscientiousness. *Human Performance, 26,* 20–43.

Hsiao, H. (2013). Anthropometric procedures for protective equipment sizing and design. *Human Factors, 55*(1), 6–35.

Huffcutt, A. I. (2011). An empirical review of the employment interview construct literature. *International Journal of Selection and Assessment, 19*(1), 62–81.

Kethley, R. B., & Terpstra, D. E. (2005). An analysis of litigation associated with the use of the application form in the selection process. *Public Personnel Management, 34*(4), 357–375.

Lee, S., & Bro, R. (2008). Regional differences in world human body dimensions: The multi-way analysis approach. *Theoretical Issues in Ergonomics Science, 9*(4), 325–345.

Levashina, J., Hartwell, C. J., Morgenson, F. P., & Campion, M. A. (2014). The structured employment interview: Narrative and quantitative review of the research literature. *Personnel Psychology, 67*(1), 241–293.

McCrae, R. R., & Costa, P. (1990). *Personality in adulthood.* New York: Guilford.

McEvily, B., Perrone, V., & Zaheer, A. (2003). Trust as an organizing principle. *Organization Science, 14,* 91–103.

McGovern, P. M., Vesley, D., Kochevar, L., Gershon, R., Rhame, F. S., & Anderson, E. (2000). Factors affecting universal precaution compliance. *Journal of Business and Psychology, 15,* 149–161.

McMullan, M., Jones, R., & Lea, S. (2010). Patient safety: Numerical skills and drug calculation abilities of nursing students and registered nurses. *Journal of Advanced Nursing, 66*(4), 891–899.

Morgeson, F. P., & Campion, M. A. (1997). Social and cognitive sources of potential inaccuracy in job analysis. *Journal of Applied Psychology, 82*(5), 627–655.

Morgeson, F. P., Campion, M. A., Dipboye, R. L., Hollenbeck, J. R., Murphy, K., & Schmitt, N. (2007). Reconsidering the use of personality tests in personnel selection contexts. *Personnel Psychology, 60*(3), 683–729.

Neal, A., & Griffin, M. A. (2004). Safety climate and safety at work. In J. Barling & M. Frone (Eds.), *The psychology of work place safety* (pp. 15–34). Washington, DC: American Psychological Association.

Pelot, R. P., Dwyer, J. W., Deakin, J. M., & McCabe, J. F. (1999). The design of a simulated forcible entry test for fire fighters. *Applied Ergonomics, 30,* 137–146.

Postlethwaite, B., Robbins, S., Rickerson, J., & McKinniss, T. (2009). The moderation of conscientiousness by cognitive ability when predicting workplace safety behavior. *Personality and Individual Differences, 47,* 711–716.

Prien, E. P., Goodstein, L. D., Goodstein, J., & Gamble, L. G. (2009). *A practical guide to job analysis.* San Francisco: Pfeiffer.

Probst, T. M., Graso, M., Estrada, A. X., & Greer, S. (2013). Consideration of future safety consequences: A new predictor of employee safety. *Accident Analysis and Prevention, 55,* 124–134.

Ramsey, J. R. (1985). Ergonomic factors in task analysis for consumer product safety. *Journal of Occupational Accidents, 7,* 113–123.

Rozenfeld, O., Sacks, R., & Rosenfeld, Y. (2009). CHASTE—construction hazard analysis with spatial and temporal exposure. *Construction Management and Economics, 27*(7), 625–638.

Rozenfeld, O., Sacks, R., Rosenfeld, Y., & Baum, H. (2010). Construction job safety analysis. *Safety Science, 48,* 491–498.

Rundmo, T. (1996). Associations between risk perception and safety. *Safety Science, 24,* 197–209.

Rundmo, T., & Hale, A. R. (2003). Managers attitudes towards safety and accidents prevention. *Safety Science, 41,* 557–574.

Salgado, J. E. (1997). The five factor model of personality and job performance in the European community. *Journal of Applied Psychology, 82,* 30–43.

Schmidt, F. L., & Hunter, J. E. (1998). The validity and utility of selection methods in personnel psychology: Practical and theoretical implications of 85 years of research findings. *Psychological Bulletin, 124*(2), 262–274.

Scroggins, W. A., Thomas, S. L., & Morris, J. A. (2009). Psychological testing in personnel selection, part III: The resurgence of personality testing. *Public Personnel Management, 38*(1), 67–77.

Simonet, S., & Wilde, G. J. S. (1997). Risk: Perception, acceptance and homeostasis. *Applied Psychology: An International Review, 46*(3), 235–252.

Smith, M. J., Cohen, H. H., Cohen, A., & Cleveland, R. J. (1978). Characteristics of successful safety programs. *Journal of Safety Research, 10*, 5–15.

Thompson, D. E., & Thompson, T. A. (1982). Court standards for job analysis in test validation. *Personnel Psychology, 35*, 865–874.

Visser, E., Pijl, Y. J., Stolk, R. P., Neeleman, J., & Rosmalen, J. G. M. (2007). Accident proneness, does it exist? A review and meta-analysis. *Accident Analysis and Prevention, 39*, 556–564.

Walker, A. (2013). Outcomes associated with breach and fulfillment of the psychological contract of safety. *Journal of Safety Research, 47*, 31–37.

Walker, A., & Hutton, D. H. (2006). The application of the psychological contract to workplace safety. *Journal of Safety Research, 37*, 433–441.

Wallace, J. C., & Vodanovich, S. J. (2003). Workplace safety performance: Conscientiousness, cognitive failure and their interaction. *Journal of Occupational Health Psychology, 8*(4), 316–327.

Wilde, G. J. S., Robertson, L. S., & Pless, I. B. (2002). For and against: Does risk homoeostasis theory have implications for road safety. *British Medical Journal, 324*, 1149–1152.

Wilson, M. (2012). *The handbook of work analysis: Methods, systems, applications and science of work measurement in organizations.* New York: Routledge.

Wood, R., & Payne, T. (1998). *Competency-based recruitment and selection.* New York: Wiley.

Chapter 6
The Influences of Socialization and Prestart Training on New Employee Safety

6.1 Introduction

This chapter discusses the section of Fig. 1.1 in Chap. 1 labeled *prestart training processes*. While there are typically legal requirements to ensure a new employee is suitably trained to perform their job, the extent of prestart training and its effectiveness will vary considerably across organizations. Research has shown that employee's perceptions of prestart training are very likely to have a negative impact on new employee safety, and the nature of this potential impact is outlined below. Before discussing training practices, the chapter briefly examines new employee socialization processes and how these might influence safety. In order to ensure that employee's perceptions of prestart training are valid, it is necessary for training to adopt best practice. As a guide to best practice, several sections in this chapter discuss how training can be designed to maximize the likelihood of an effective outcome. This chapter concludes with a discussion of a strategy which can be used to ensure that employees do not form incorrect assumptions about the effectiveness of socialization processes and prestart training.

6.2 Employees Perception of Prestart Training: The Dangers

Individuals that undertake training can potentially not only learn from the training, but also make assumptions that other employees are learning, have learnt, or will learn from the same training. Given the fact that training does not always deliver the desired outcome, and there will be large individual differences in training outcomes, a general assumption that training will achieve its objectives is very risky. That is,

© Springer International Publishing Switzerland 2015
C.D.B. Burt, *New Employee Safety*,
DOI 10.1007/978-3-319-18684-9_6

trusting organizational prestart training processes to deliver new employees who will work safely can potentially be dangerous. Furthermore, research strongly suggests that positive perceptions of training effectiveness can potentially suppress employee behaviors on the job which might be needed to ensure new employee safety.

Three studies that I conducted with colleagues (e.g., Burt et al. 2009; Burt and Stevenson 2009; Burt and Hislop 2013) questioned employees about prestart training in their organization (see Chap. 9 for the scale items, and psychometric data on the scales used). The key focus of the research was the degree to which employees trusted selection processes and prestart training to help deliver a new employee that would work safely. The studies also measured employees trust in new employees to work safely, their perception of the degree of safety risk associated with new employees, and the extent to which the employees (participants) engaged in safety promoting or ensuring behaviors with new employees (referred to as compensatory behaviors). Table 6.1 provides a summary of the samples examined in the studies, and the key findings. Figure 6.1 provides an interpretation of the pattern of correlations shown in Table 6.1. Basically, employees in the three studies trusted their organization's prestart training to deliver an employee that would work safely, reduced their perceived risk in new employees as their trust in prestart training increased, and reduced their compensatory behaviors toward new employees as the perceived risk from new employees decreased.

While the research only reported correlations between the variables, and a degree of caution is required in translating correlations into a causal pattern, it does seem logical to interpret the relationship in the way shown in Fig. 6.1. It would also appear that the assumed causal pathway is not unique to a particular sample, as one

Table 6.1 Relationships between perceptions of prestart training, trust, risk-taking, and compensatory behaviors

Study	Sample	Correlation between trust in prestart training processes and trust in new employees to work safely	Correlation between trust in new employees and perceived safety risk from new employees	Correlation between perceived safety risk from new employees and compensatory behaviors to help ensure new employee safety
Burt et al. (2009)	128 forestry workers	0.22*	−0.20**	0.33*
Burt and Stevenson (2009)	154 professional firefighters	0.50*	−0.24**	0.43*
Burt and Hislop (2013)	118 employees in high-risk jobs from 5 organizations	0.28*	−0.13	0.43*

$*P < 0.01$; $**P < 0.05$

Fig. 6.1 Causal
interpretation of the results
shown in Table 6.1

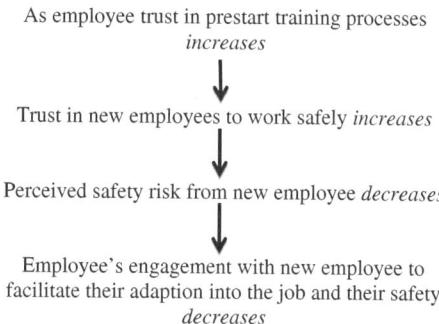

As employee trust in prestart training processes
increases

↓

Trust in new employees to work safely *increases*

↓

Perceived safety risk from new employee *decreases*

↓

Employee's engagement with new employee to
facilitate their adaption into the job and their safety
decreases

of the most striking factors evident in the results shown in Table 6.1 is the consistency of the findings across the 3 studies which sampled from different industries. It appears that across different industries and organizations, there was a reasonably consistent positive perception of prestart training, and this was associated with how employees felt about new employees (trust and risk) and their behavior toward the new employees. There appears to be two options for how organizations can respond to these results. The options are not either or, but rather should both be adopted. First, organizations should attempt to ensure prestart training is in fact effective. Secondly, organizations can attempt to ensure employees are informed about the limitations of training, and how unfounded assumptions about training effectiveness may put new employees, and themselves, at risk.

It is also worth noting that the research findings of Burt et al. (2009), Burt and Stevenson (2009), and Burt and Hislop (2013) are consistent with other research which has examined how employee's risk perceptions are related to safety. Leiter et al. (2009) examined 350 workers' perceptions of the control they had gained over hazards from their safety training and how this was associated with their risk perceptions. The study found that as perceptions of control increased, perceptions of risk decreased. It is easy to see how employees can extend this relationship to new employees. All that is required is the assumption that others (new employees) will gain the same level of control from safety training as they gained and that this will allow the new employee to work in a safer way.

The research findings in relationship to training shown in Table 6.1 can also be explained by Wilde's risk homoeostasis theory (RHT) (see Glendon et al. 1996; Wilde et al. 2002; Simonet and Wilde 1997). RHT predicts that as safety features are added to a system, individuals will increase their risk-taking. It is easy to see how perceptions that prestart training will make a new employee work more safely can be associated with a reduction in perceived risk from the new employee, and result in employees taking more risks around them than are justified by their status as a new employee. While the research reported above (e.g., Burt et al. 2009; Burt and Stevenson 2009; Burt and Hislop 2013) did not question employees about socialization processes, it seems reasonable to predict that similar attitudes might form about socialization processes. As such, the next section briefly examines research on new employee socialization.

6.3 Socialization Processes

For the sake of clarity, I assume that socialization processes provide information (e.g., general safety rules, evacuation procedures) which an organization would like all new employees to have (irrespective of their job), whereas prestart training provides information, skills, etc., which are specifically needed for the new employee's job. To ensure new employee's safety, organizations need to adopt best practice in terms of both socialization and prestart training processes. Furthermore, they need to understand how individual differences across a range of variables will influence the outcome of these processes. It is also essential that, based on evaluation evidence, the organization communicates to employees the risks associated with making assumptions about the effectiveness of socialization and training processes. In this section, I will examine the literature on socialization tactics, as well as important individual difference variables which need to be taken into account.

Socialization (and prestart training) is in its broadest sense aimed at facilitating new employee adjustment. Feldman (1981) suggested that new employee adjustment includes 3 components: *resolution of role demands*, such as understanding the job tasks to be performed, task prioritization, and time allocation factors which collectively lead to role clarity; *task mastery* which involves learning tasks and gaining confidence in the role; and *adjustment to one's group* defined as feeling liked and accepted by co-workers. Resolution of role demands and task mastery should involve an aspect of training. *Adjustment to one's group* can be facilitated by the organization, but also involves a number of other psychological processes and will be dealt with in detail in Chap. 7. In addition to the aspects of adjustment noted by Feldman (1981) which are primarily linked to the new employee's job and colleagues, new employees need to adjust to the organization's policies and general procedures. Part of this is adjusting to the way that the organization manages health and safety.

As noted, socialization processes are likely to be perceived in a similar way to prestart training. That is, if an organization has a socialization process where new employees are introduced to safety policy and procedures, it might be reasonable to assume that this will have a positive impact on the new employee's safety-related behavior on the job. A study by Mullen (2004) supported this proposition, finding that early socialization processes could have a positive influence on safety behavior. Of course, socialization processes may have no effect at all. A new employee, who is asked during socialization to learn the organization's safety policy and procedures, understand the organization's emphasis on safety (its safety culture in the form of norms, beliefs, roles, attitudes, and practices), and learn how to complete appropriate forms (such as hazard sheets, near miss reports), may simply not achieve these expected outcomes. To help increase the chances that socialization will have a positive impact on new employees' safety, best practice should be adopted.

Bauer et al. (2007) provide a useful review of studies that have examined new employee adjustment from organizational socialization processes. While the review did not focus on safety, and in fact, there appears to be little research which has directly addressed the relationship between socialization processes and safety, it provides a lot of useful information about both individual difference characteristics and process characteristics which organizations can use to guide the development of their safety-specific (and general) socialization processes.

A key dimension associated with the success of socialization processes is the *information seeking* characteristics of the new employee (Van Maanen and Schein 1979). Individuals seek information in order to create predictability. However, individuals will vary in the degree to which they seek information, the degree to which they wish to reduce uncertainty, and the degree to which they require predictability. However, it is not easy to predict an individual's information seeking behavior as it is associated with a large number of factors. Research has found that information seeking increases with higher self-esteem (e.g., Weiss 1977), lower tolerance for ambiguity (e.g., Norton 1975), higher cognitive (integrative) complexity (Stabell 1978), higher self-efficacy (Jones 1986), the trait of specific curiosity (Harrison et al. 2011), more experience in making role transitions, and increased familiarity with the new work environment (Pavelchak et al. 1986).

Information seeking can also be motivated by the psychological contract which forms between a new employee and their employing organization (De Vos et al. 2005). Psychological contract theory is based on the concept of reciprocity where the employee's work is given in exchange for terms and conditions (such as pay, recognition) (Rousseau 1995). The psychological contract begins at the time of recruitment, and organizations make promises and delivers expectations at every step during the recruitment and selection process. New employees can be motivated to seek information during socialization to help confirm that the organization is keeping its side of the psychological contract. Furthermore, the new employee is expecting to hear and see things during socialization which are consistent with what they were told during earlier stages of the entry process.

The individual difference variables which are associated with information seeking create a dilemma in that employees cannot assume that every new employee is or was seeking to gain the same information from socialization, as to what they themselves gained. Given the complexity of the factors which seem to predict information seeking, it is perhaps easiest to rely on a measure of the success of socialization as a way of handling these individual differences. If a particular new employee does not meet the required assessment standard, they would be asked to reenter the socialization process, with the aim of ensuring all new employees, regardless of their individual information seeking behavior, achieve the same outcome from the socialization process.

Appropriate structuring of the socialization process itself should help ensure new employees have the best chance to gain the required knowledge. Van Maanen and Schein (1979) suggest 6 ways in which socialization processes can vary. Table 6.2 lists the 6 dimensions and also notes the implications for employees' assumptions about the effectiveness of socialization. I have also added two other dimensions (facilitator training and evaluation) which I think can potentially have an impact on the outcome of socialization. Inspection of the process factors shown in Table 6.2

Table 6.2 Aspects of the socialization process and their impact on *learning variability* across new employees

Socialization processes	Potential learning variability across new employees
Collective: newcomers common experiences as part of group	Less
Versus	
Individual: newcomers separate experiences	More
Formal: newcomer segregated from other and off the job	Less
Versus	
Informal: newcomer with employees and on the job	More
Sequential: newcomer knows what phases they need to go through	Less
Versus	
Random: the progression is more ambiguous	More
Fixed: timetable of when socialization is completed	Less
Versus	
Variable: no specific timetable	More
Serial: completed with the help of role model or employee	Less
Versus	
Disjunctive: without help	More
Investiture: new employee receives feedback	Less
Versus	
Divestiture: no feedback	More
Person responsible for delivering socialization is a specialist	Less
Versus	
Person responsible for delivering socialization is a generalist	More
Assessment standard Yes: specific level required for passing	Less
Versus	
No assessment is conducted	More

suggests that they can be classified as *content, context* and *social* aspects (Jones 1986). The safety-specific content of the socialization process needs to be carefully designed and should address the organization's safety culture, safety climate factors, and all the procedural steps which the organization has in place to manage safety. As shown in Table 6.2, the context aspect needs to be formal with a focus on being systematic. The *social* component of the socialization process provides an avenue through which the new employee's integration into their work group can begin. As noted above, this aspect of socialization is discussed in more detail in Chap. 7.

While the design of the socialization process is going to have a significant impact on the process's success, it is vital that evaluation of the socialization process is undertaken. The evaluation focus should be to ensure that every new employee completes the process with the required standard of knowledge. Any variation in knowledge is likely to lead to variation in behavior once in the job and, as noted above, can potentially lead to employees' assumptions about what a new employee learnt during the socialization being erroneous. Furthermore, the organization needs to ensure that new employees realize that if they do not achieve the required standard in the post-socialization evaluation, they will be asked to retake the socialization process. Clearly, the organization has to commit resources to the evaluation process, and be willing to allow the new employees the time to retake socialization if they fail the evaluation. However, a total commitment to new employee safety requires this.

6.4 Prestart Training

Prestart training should include components of both safety training and job-specific training, and must occur before the new employee begins their job. On-the-job training may also occur and may cover similar topics, but the extremely high accident rate associated with new employees needs to be addressed by prestart training. It is tempting to think that every new employee will receive some sort of prestart training. However, and despite the existence of a legal obligation to do so in many countries, the evidence suggests this is far from the case. For example, a Canadian study of 28,639 workers found that only one in five employees (21.4 %) received safety training in their first year of employment, and there was no evidence that training was being targeted at high-risk groups such as younger workers (Smith and Mustard 2007). Given the analysis considered the first year of employment, the proportion receiving training before they started work was undoubtedly much smaller than 1 in 5.

From both a safety, and a performance perspective, having training is better than not having training, and a lack of training has been noted as a contributing factor in studies of accidents (e.g., Burke et al. 2011; Wagenaar and Groeneweg 1987). In contrast, studies have shown that training can improve safety attitudes (e.g., DeJoy et al. 2000; Harvey et al. 2001) and decrease lost-time accidents (e.g., Harshbarger and Rose 1991; Vredenburgh 2002). However, training is not always effective (Clemes et al. 2010; Bell and Grushecky 2006; Laberge et al. 2014), learning from training is not always applied in the work environment (Clemes et al. 2010), and inadequate or inappropriate safety training has been identified as a cause of accidents (Crowe 1985; Holman et al. 1987; MacFarlane 1979). Furthermore, while ineffective training may not directly cause an accident, the research discussed above on employees' perceptions of training (e.g., Burt et al. 2009; Burt and Stevenson 2009; Burt and Hislop 2013) suggests that ineffective training may decrease workplace safety through a series of flawed associations. Therefore, it is vital that every attempt is made to ensure that training is effective.

To be effective, a safety training program needs to be characterized by a number of features. First, a needs assessment or analysis should be undertaken to identify the content of the training program. Content should be both specific to the new employee's job and guided by research findings in the safety literature. The presentation of the content should follow a systematic training model and adopt best practice methods, which should also allow for a consideration of new employee individual differences. New employees that complete prestart training should be required to meet an assessment standard before they begin work. Extensive evaluation of the training's effectiveness and post-training transfer evaluations should be undertaken. Finally, the training program's ability to deliver new employees that meet the required standards should be clearly communicated to employees. Each of these components of an effective training program is expanded on in the following sections.

6.4.1 The Needs Assessment: Training Program Content

Organizations should have priority areas which have been identified from the analysis of previous accidents and near misses which form part of the contents of a training program. Furthermore, the specific business or activity which the organization is engaged in will occasion the need for specific aspects of training. Depending on the specific business, and job activities, it may be possible to find very extensive and useful material which can be used to develop training program content. For example, there is an extensive literature on manual handling training: Clemes et al. (2010) review 53 papers which examined the effectiveness of manual handling training. Clearly, there will be key points in this literature which can be applied to the development of a manual handling training program.

There is also likely to be a need for prestart safety training to cover general topics such as the use of safety equipment, use of fire-fighting equipment, evacuation procedures, and first aid. Reese (2001) noted the importance of providing extensive prestart training, linking its importance to the high accident rate associated with new employees, and provided a list of topics which should be covered: 'accident reporting procedures, basic hazard identification and reporting, chemical safety, company's basic philosophy on safety and health, company's safety and health rules, confined space entry, electrical safety, emergency response procedures (fire, spill, etc.), eyewash and shower locations, fall protection, fire prevention and protection, first aid/CPR, hand tool safety, hazard communications, housekeeping, injury reporting procedures, ladder safety, lockout/tagout procedures, machine guarding, machine safety, material handling, mobile equipment, medical facility location, personal responsibility for safety, rules regarding dress code, conduct and expectations, unsafe acts/conditions reporting procedures, use of personal protective equipment (PPE)' (p. 231).

The importance of prestart training having a focus on the recognition of hazards and hazardous actions cannot be overstated. New employees need to be skilled in risk appraisal or perception, which is the ability to recognize a hazard's ability to harm and to estimate the probability of being harmed (Cox and Tait 1991). Simply being told to watch out for hazards is unlikely to help new employees stay safe. Research has found that *perceived danger* (perceptions that a risk exists) increases safety compliance (e.g., Vredenburgh and Cohen 1995), and recognition of hazards and hazardous actions provides for the perception of danger.

6.4.2 Training Program Structure and Methods

There are many books (e.g., ReVelle and Stephenson 1995) and research papers (e.g., Martin et al. 2014) on training methods. Martin et al. (2014) provides a description of 13 different training methods: case study, games-based training, internship, job rotation, job shadowing, lecture, mentoring and apprenticeship, programmed instruction, role-modeling, role play, simulation, stimulus-based training, and team-training. There is not sufficient space in this chapter to review all of these methods in detail. However, it seems clear that training which requires *active participation* is likely to be the more effective than passive methods. Burke et al. (2006) compared safety and health training methods across 95 studies (n = 20991), classifying them as using 'least engaging (lecture, pamphlets, videos), moderately engaging (programmed instruction, feedback interventions), and most engaging (training in behavioral modeling, hands-on-training)' (p. 315) methods, and found that knowledge acquisition increased and safety outcomes improved as the engagement level of the training methods increased. Clearly, a key principal is to include behavioral modeling, practice, and dialogue to maximize the engagement level of the training method, what Martin et al. (2014) refer to as *learning by doing*. However, given the high safety risk level associated with new employees, it is essential that organizations do not *exclusively* adopt training methods which involve *learning by doing*, or on-the-job learning (e.g., internship, job rotation). There is no rule which says an accident cannot happen during training.

There is ample research evidence showing that matching the training method to the trainees learning style(s) facilitates learning (Martin et al. 2014). Individuals vary in terms of the learning modality they prefer: Learning by doing, by seeing, by hearing, and considering this in the design of the training can have benefits. Other research has shown that the personality facet *trainability* can be positively associated with safety performance (Hogan and Foster 2013), suggesting yet another dimension which needs to be taken into account. Of course, it is one thing to show these finding in research papers, whereas in the real world, there are practical considerations in terms of the cost involved in potentially having to reshaping a training approach for cohorts of new employees that vary in terms of personality or preferred learning style.

Despite the practical considerations, organizations must recognize that how their training programs are designed; that is, the learning approach which is adopted is a critical aspect in the likely success of the training. Given the extensive evidence which suggests traditional training approaches may be failing, it may be time to look closely at the learning strategies incorporated into training (Laberge et al. 2014). Laberge et al. (2014) specifically focus on understanding the learning processes of young workers and how these can be used to improve health and safety training. Consistent with previous research, evidence was found to support the value of *learning* via *doing* as being superior to learning via being told. Clearly, incorporating *doing* or *activity* components into a training program is essential for a successful outcome, but at the same time, there needs to be control over the context to ensure the relevant skills and competencies are gained, while the risks of injury are removed.

6.4.3 Integration into Existing Knowledge

Two further factors require noting. First, the degree of existing knowledge that an individual has about a domain will influence their ability to learn new information. That is, new employees in the categories discussed in Chap. 3 (school leaver, career transition, occupational focused, or career focused) will bring different levels of knowledge to a training program. Cognitive psychology research has repeatedly shown how new information is integrated into existing knowledge structures, and this process allows for easier learning of information and better acquisition of skills and competencies. Therefore, it would be a mistake to 'pitch' a training program at a level which assumes more preexisting knowledge than is held by the least experienced member of the group being trained. While starting at the beginning may be repetitious for new employees with more work experience, it will ensure those with less experience have the best opportunity to acquire the knowledge and skills. It would also be a mistake to excuse a new employee who comes with years of previous work from the training.

6.4.4 Training Evaluation

Without doubt, there are many prestart and safety training programs which are simply assumed to be effective, rather than being empirically evaluated and shown to be effective. Without a systematic evaluation, a safety training program cannot be considered effective. Research on individuals responsible for the evaluation of safety training also suggests that many evaluations are not adequately designed to a standard which allows for valid conclusions as to effectiveness to be made, and there may even be occasions when the conclusion is that training is effective, when it is not effective (Vojtecky and Schmitz 1986).

The research evidence around the evaluation of safety training programs is actually rather alarming. Goldenhar et al. (2001) examined safety training in the construction sector and found that 37 % of the 45 companies examined did not quantitatively evaluate their training. Furthermore, the evaluation work undertaken by the remaining companies was not comprehensive. Bell and Grushecky (2006) claim that 'Logger safety training programs are rarely, it ever, evaluated.' (p. 53), yet there is ample evidence showing that forestry work is extremely dangerous (and that forestry workers place a lot of trust in training, e.g., Burt et al. 2009). Other industries also suffer from training evaluation issues. For example, Egan et al. (2007) describe evaluation issues associated with food safety and food hygiene training.

Training evaluation involves two key considerations: the design of the evaluation and the evaluation criteria. The design of the evaluation needs to provide for valid conclusions to be drawn. However, the most rigorous evaluation designs involve control groups which bring up ethical questions when the training has safety implications. Organizations are advised to consult the literature on training program evaluation designs (e.g., Arthur et al. 2003; Ford et al. 2010; Sackett and Mullen 1993) and select the best possible design within the limits of the context and material to be trained.

Considerable research attention has been focused on the criteria used to evaluate training programs. For many years, the four levels of training evaluation criteria developed by Kirkpatrick (i.e., reactions, learning, behavior, and results) were considered the gold standard (Alliger and Janak 1989). Research on the relationships between these criteria has found particularly weak relationships between trainee reactions and the other criteria (e.g., Alliger et al. 1997), indicating it is very unwise to assume that if trainees give a training program favorable ratings that they will have learnt anything from the training. While Kirkpatrick's work was clearly important and influential, it is now recognized that evaluation criteria can be more complex than his four-level approach. Kraiger et al. (1993) provide a useful description of a more complex evaluation model which includes cognitive outcomes (e.g., knowledge and strategies), skill-based outcomes (e.g., proceduralization, automaticity), and affective outcomes (e.g., attitudinal, motivational). Clearly, more complex evaluation criteria allow for more information to be collected about the effectiveness of a training program, and thus, more information is available to be passed on to employees so that they can form valid opinions about training.

6.4.5 Training Transfer and Maintenance

A particularly important aspect of training evaluation is determining the degree to which the knowledge, skills, and behaviors acquired during the training transfers on to the job and are maintained over time. Measures showing knowledge and skills have been learnt during socialization, and training activities do not necessarily guarantee that the transfer of these knowledge and skills into the workplace will occur (e.g., Hogan et al. 2014). Furthermore, if there is transfer from the socialization and/or training context onto the job, there is no guarantee that there will be long-term maintenance of the knowledge and/or behavior. There is an extensive literature dealing with the transfer of knowledge, skills, and behaviors from a training context into the job context (e.g., Blume et al. 2010; Burke and Hutchins 2007; Cheng and Hampson 2008). Factors associated with the trainee, the design of the training, and the organization will have an impact on motivation to transfer training to the workplace (Gegenfurtner et al. 2009a, b).

One of the most important aspects which influences training transfer is the *transfer climate*. This is basically the environment into which the trained knowledge, behavior, and skills are taken. Smith-Crowe et al. (2003) showed how the transfer climate (both general organizational climate factors and safety-specific climate factors) can moderate the relationship between safety knowledge delivered though training and safety outcomes. Burke et al. (2008) also found that safety climate moderated the transfer of safety training and additionally showed that a dimension of national culture (uncertainty avoidance) was a moderator of the safety training transfer. Put simply, if the transfer climate is not supportive (including manager and peer support), knowledge and behavior learnt during socialization and training will not be used on the job or, if used initially on the job, will be quickly lost due to the lack of support (Colquitt et al. 2000).

6.5 Communicating Realistic Socialization and Training Outcome Expectations

It is unclear how many organizations attempt to communicate to their employees' realistic expectations about the effectiveness of socialization and prestart training processes. Like many organizational processes, socialization and prestart training may simply be lorded as positive features of the organizations human resource management services, whether or not this is justified by valid evaluation results. Arguably considerable safety advantage for new employees (and employees in general) may be gained by encouraging employees to be skeptical about the impact of socialization and prestart training. Furthermore, even where evaluation results are showing positive outcomes, individual differences which can moderate outcome effects need to be taken into consideration. That is, while the last new employee may have benefited significantly from a prestart training program, the next may not benefit as much. If co-workers assume similar outcomes are always achieved, then they may face risks which they are not expecting.

Another important issue to communicate in relation to prestart training is that whereas safety training may attempt to ensure safety, new team members will still lack familiarity (see Chap. 7) with the specific equipment used by the team, their specific work environment, and the specific way members of the team do their job. Thus, prestart training generally, and always, has a limited potential to ensure new employee safety.

6.6 Conclusions

Research has repeatedly shown that socialization and training processes can vary considerably in their effectiveness. Literature on both socialization and training offers many suggestions about how these processes can be designed to maximize the possibility of positive outcomes. Research also suggests that employees, at least in the 3 studies cited in Table 6.1, seem to generally have a very optimistic view of prestart training, and on average rate, it is likely to have a positive impact on new employee safety. This trust in training was found to be associated with a decrease in perceived risk from new employees, and an associated decrease in employees' engagement in behaviors which will help ensure a new employee's safety and help them adapt to the job. Clearly, if socialization and prestart training are not working correctly, are not achieving their objectives, and employees think they are, negative safety outcomes are likely. In its simplest terms, the message from this chapter is that employees should be encouraged to be skeptical about the ability of socialization and prestart training to help ensure new employee safety: Employees should not let trust in organizational processes stop them from being careful around new employees, and from helping to ensure their safety.

References

Alliger, G. M., & Janak, E. A. (1989). Kirkpatrick's levels of training criteria: Thirty years later. *Personnel Psychology, 42*, 331–342.

Alliger, G. M., Tannenbaum, S. I., Bennett, W, Jr, Traver, H., & Shotland, A. (1997). A meta-analysis of the relations among training criteria. *Personnel Psychology, 50*, 341–358.

Arthur, W., Bennett, W., Edens, P. S., & Bell, S. T. (2003). Effectiveness of training in organizations: A meta-analysis of design and evaluation features. *Journal of Applied Psychology, 88*(2), 234–245.

Bauer, T. N., Bodner, T., Erdogan, B., Truxillo, D. M., & Tucker, J. S. (2007). Newcomer adjustment during organizational socialization: A meta-analysis review of antecedents, outcomes, and methods. *Journal of Applied Psychology, 92*(3), 707–721.

Bell, J. L., & Grushecky, S. T. (2006). Evaluating the effectiveness of a logger safety training program. *Journal of Safety Research, 37*, 53–61.

Blume, B. D., Ford, J. K., Baldwin, T. T., & Huang, J. L. (2010). Transfer of training: A meta-analytic review. *Journal of Management, 36*, 1065–1105.

Burke, M. J., Chan-Serafin, S., Salvador, R., Smith, A., & Sarpy, S. A. (2008). The role of national culture and organizational climate in safety training effectiveness. *European Journal of Work and Organizational Psychology, 17*(1), 133–152.

Burke, L. A., & Hutchins, H. M. (2007). Training transfer: An integrative literature review. *Human Resource Development Review, 6*, 263–296.

Burke, M. J., Salvador, R. O., Smith-Crowe, K., Chan-Serafin, S., Smith, A., & Sonesh, S. (2011). The dread factor: How hazards and safety training influence learning and performance. *Journal of Applied Psychology, 96*(1), 46–70.

Burke, M. J., Sarpy, S. A., Smith-Crowe, K., Chan-Serafin, S., Salvador, R. O., & Islam, G. (2006). Relative effectiveness of worker safety and health training methods. *American Journal of Public Health, 96*(2), 315–324.

Burt, C. D. B., Chmiel, N., & Hayes, P. (2009). Implications of turnover for safety attitudes and behaviour in work teams. *Safety Science, 47*, 1002–1006.

Burt, C. D. B., & Hislop, H. (2013). Developing safety specific trust in new recruits: The dilemma and a possible solution. *Journal of Health, Safety and Environment, 29*(3), 161–173.

Burt, C. D. B., & Stevenson, R. J. (2009). The relationship between recruitment processes, familiarity, trust, perceived risk and safety. *Journal of Safety Research, 40*, 365–369.

Cheng, E. W. L., & Hampson, I. (2008). Transfer of training: A review and new insights. *International Journal of Management Reviews, 10*, 327–341.

Clemes, S. A., Haslam, C. O., & Haslam, R. A. (2010). What constitutes effective manual handling training? A systematic review. *Occupational Medicine, 60*, 101–107.

Colquitt, J. A., Le Pine, J. A., & Noe, R. A. (2000). Towards an integrative theory of training motivation: A meta-analytic path analysis of 20 years of research. *Journal of Applied Psychology, 85*(5), 678–707.

Cox, S., & Tait, R. (1991). *Reliability, safety and the human factor*. Stoneham, MA: Butterworth-Heineman.

Crowe, M. P. (1985). Felling techniques in Australian hardwood forests. *Australian Forestry, 48* (2), 84–94.

Dejoy, D. M., Searcy, C. A., Murphy, L. R. & Gershon, R. R. M. (2000). Behavior-diagnostic analysis of compliance with universal precautions among nurses. *Journal of Occupational Health Psychology, 5*(1), 127–141.

De Vos, A., Buyens, D., & Schalk, R. (2005). Making sense of a new employment relationship: Psychological contract-related information seeking and the role of work values and locus of control. *International Journal of Selection and Assessment, 13*(1), 41–52.

Egan, M. B., Raats, M. M., Grubb, S. M., Eves, A., Lumbers, M. L., Dean, M. S., & Adams, M. R. (2007). A review of food safety and food hygiene training studies in the commercial sector. *Food Control, 18*, 1180–1190.

Feldman, D. C. (1981). The multiple socialization of organization members. *Academy of Management Review, 6*, 309–318.

Ford, J. K., Kraiger, K., & Merritt, S. M. (2010). An updated review of the multidimensionality of training outcomes: New directions for training evaluation research. In S. W. J. Kozlowski & E. Salas (Eds.), *Learning, training and development in organizations* (pp. 135–165). NY: Routledge/Taylor & Francis Group.

Gegenfurtner, A., Festner, D., Gallenberger, W., Lehtinen, E., & Gruber, H. (2009a). Predicting autonomous and controlled motivation to transfer training. *International Journal of Training and Development, 13*(2), 124–138.

Gegenfurtner, A., Veermans, K., Festner, D., & Gruber, H. (2009b). Motivation to transfer training: An integrative literature review. *Human Resource Development Review, 8*, 403–423.

Glendon, A. I., Hoyes, T. W., Haigney, D. E., & Taylor, R. G. (1996). A review of risk homeostasis theory in simulated environments. *Safety Science, 22*, 15–25.

Goldenhar, L. M., Moran, S. K., & Colligan, M. (2001). Health and safety training in a sample of open-shop construction companies. *Journal of Safety Research, 32*, 237–252.

Harrison, S. H., Sluss, D. M., & Ashforth, B. E. (2011). Curiosity adapted the cat: The role of trait curiosity in newcomer adaption. *Journal of Applied Psychology, 96*(1), 211–220.

Hogan, J., & Foster, J. (2013). Multifaceted personality predictors of workplace safety performance: More than conscientiousness. *Human Performance, 26*, 20–43.

Hogan, D. A. M., Greiner, B. A., & O'Sullivan, L. (2014). The effect of manual handling training on achieving training transfer, employee's behavior change and subsequent reduction of work-related musculoskeletal disorders: A systematic review. *Ergonomics, 57*(1), 93–107.

Holman, R. G., Olszewski, A., & Maier, R. V. (1987). The epidemiology of logging injuries in the Northwest. *Journal of Trauma, 27*(9), 1044–1050.

Harshbarger, D., & Rose, T. (1991). New possibilities in safety performance and the control of workers compensation costs. *Journal of Occupational Rehabilitation, 1*, 133–143.

Harvey, J., Bolam, H. D., Gregory, D., & Erdos, G. (2001). The effectiveness of training to change safety culture and attitudes within a highly regulated environment. *Personnel Review, 30*, 615–646.

Jones, G. R. (1986). Socialization tactics, self-efficacy, and newcomers adjustment to organizations. *Academy of Management Journal, 29*, 262–279.

Laberge, M., MacEachen, E., & Calvet, B. (2014). Why are occupational health and safety training approaches not effective? Understanding young workers learning processes using an ergonomic lens. *Safety Science, 68*, 250–257.

Leiter, M. P., Zanaletti, W., & Argentero, P. (2009). Occupational risk perceptions, safety training and injury prevention: Testing a model in the Italian printing industry. *Journal of Occupational Health Psychology, 14*(1), 1–10.

Kraiger, K., Ford, J. K., & Salas, E. (1993). Application of cognitive, skill-based, and affective theories of learning outcomes to new methods of training evaluation. *Journal of Applied Psychology, 78*, 311–328.

MacFarlane, I. (1979). Injuries and death in forestry: A NZ experience. *Logging Industry Research Association Technical Release, 1*(1), 1–4.

Martin, B. O., Kolomitro, K., & Lam, T. C. M. (2014). Training methods: A review and analysis. *Human Resource Development Review, 13*(1), 11–35.

Mullen, J. (2004). Investigating factors that influence individual safety behavior at work. *Journal of Safety Research, 35*, 275–285.

Norton, R. W. (1975). Measurement of ambiguity tolerances. *Journal of Personality Assessment, 39*, 607–619.

Pavelchak, M. A., Moreland, R. L., & Levine, J. M. (1986). Effects of prior group membership on subsequent reconnaissance activities. *Journal of Personality and Social Psychology, 50*, 56–66.

Reese, C. D. (2001). *Accident/incident prevention techniques.* London: Taylor & Francis.

ReVelle, J. B., & Stephenson, J. (1995). *Safety training methods: Practical solutions for the next millennium.* New York: Wiley.

Rousseau, D. M. (1995). *Psychological contracts in organizations: Understanding written and unwritten agreements.* Thousand Oaks, CA: Sage.

Sackett, P. R., & Mullen, E. J. (1993). Beyond formal experimental design: Towards an expanded view of the training evaluation process. *Personnel Psychology, 46*, 613–627.

Simonet, S., & Wilde, G. J. S. (1997). Risk: Perception, acceptance and Homeostasis. *Applied Psychology: An International Review, 46*(3), 235–252.

Smith, P. M., & Mustard, C. A. (2007). How many employees receive safety training during their first year of a new job? *Injury Prevention, 13*, 37–41.

Smith-Crowe, K., Burke, M. J., & Landis, R. (2003). Organizational climate as a moderator of safety knowledge-safety performance relationships. *Journal of Organizational Behavior, 24*, 861–876.

Stabell, C. B. (1978). Integrative complexity of information environment perception and information use: An empirical investigation. *Organizational Behavior and Human Performance, 22*, 116–142.

Van Maanen, J., & Schein, E. H. (1979). Towards a theory of organizational socialization. *Research in Organizational Behavior, 1*, 209–264.

Vojtecky, M. A., & Schmitz, M. F. (1986). Program evaluation and health and safety training. *Journal of Safety Research, 17*, 57–63.

Vredenburgh, A. G. (2002). Organizational safety: Which management practices are most effective in reducing employee injury rates? *Journal of Safety Research, 33*, 259–276.

Vredenburgh, A. G., & Cohen, H. H. (1995). High-risk recreational activities: Skiing and scuba—what predicts compliance with warnings. *International Journal of Industrial Ergonomics, 15*, 123–128.

Wagenaar, W. A., & Groeneweg, J. (1987). Accidents at sea: Multiple causes and impossible consequences. *International Journal of Man-Machine Studies, 27*, 587–598.

Weiss, H. M. (1977). Subordinate imitation of supervisor behavior: The role of modeling in organizational socialization. *Organizational Behavior and Human Performance, 19*, 89–105.

Wilde, G. J. S., Robertson, L. S., & Pless, I. B. (2002). For and against: Does risk homoeostasis theory have implications for road safety? *British Medical Journal, 324*, 1149–1152.

Chapter 7
The Initial Employment Period: Behaviors, Familiarity, Adaptation and Trust Development

7.1 Introduction

Figure 1.1 (in Chap. 1) shows a box titled *familiarization, adaption and trust development*, and this chapter deals with these three processes which begin on the day the new employee starts work (remembering of course that the new employee should have been through socialization and prestart training processes before they actually start working (see Chap. 6)). Two other parts of Fig. 1.1 are also discussed, these being factors associated with task assignment, and issues around the provision of supervision, and how this is likely to change. This is followed by a discussion of co-workers behavior towards new employees. All of these factors and processes can positively influence new employees familiarity their new job, the work environment, and with their co-workers, and ultimately their adaption to the job, work environment and co-workers. During the initial 3 month period of employment, it is also likely that the new employees behavior will change. Each of the processes noted above have links with the development of trust between employees, and between employees and management. Thus the chapter concludes with a discussion of trust, how trust develops in the initial period of employment, and a process which can be used to build trust in a *safety enhancing* way. The overall aim is to identify aspects of each process which can be carefully managed in order to enhance new employee safety.

7.2 The Adaptation Process

Several researchers (e.g., Ashford and Black 1996; Bauer and Green 1994; Chan and Schmitt 2000) define the initial period of employment in a job as the *adaptation process*. Others have referred to it as the *encounter period*, and describe it as characterized by uncertainty (e.g., Miller and Jablin 1991). During this initial

© Springer International Publishing Switzerland 2015
C.D.B. Burt, *New Employee Safety*,
DOI 10.1007/978-3-319-18684-9_7

period, and prior to adaptation, new employees are very likely to be a safety risk to them self, and to those they work with. While it is hard to put a specific time frame on how long it takes for a new employee to fully adapt, it is reasonable to assume that adaption processes are happening for at least the initial 3 months of employment. Clearly there is a lot happening in the initial period of employment in a new job. For example, it is a time when the new employee is getting to know his or her co-workers, developing an understanding how tasks are performed, and developing a relationship with their supervisor. These extra aspects of work are primarily associated with the initial period of employment. Furthermore, each of the processes associated with these aspects requires a degree of attention and effort on the part of the new employee. This is attention and effort which is from a limited resource, and will take away from the attention and effort which the new employee can devote to their job, and can devote to maintaining *situational awareness*.

Situational awareness involves the perception of the environment that an individual is in, the comprehension of its meaning, and the projection of the individuals status into the near future (Endsley 1995). A new employee not only faces the processes required to adapt to the job, co-workers and management, but to ensure their safety, they must also maintain situational awareness. In other words a new employee must know and understand what is going on around them while they adapt. Arguably, developing new employee adaption processes which are effective and efficient should help improve both the outcomes of the processes, and the new employee's ability to maintain situational awareness during their adaption.

7.3 Task Assignment

It is well established that work groups establish their own social order. Furthermore, this social order can be associated with the development of work group norms. Part of a work group's normative structure, maybe an understanding that the 'new guy' (new employee) is required to do certain tasks (also see Chap. 4, Sect. 4.2.3). Unfortunately, these tasks are likely to be the tasks which more senior members of the work group do not want to do because they are for example, dirty, monotonous, and/or risky. If this situation does prevail within a work group it potentially creates an *inequality of risk exposure* across employees with different job tenure. New employees with short job tenure are thus exposed to more risk, potentially resulting in higher accident rates. More senior employees on the other hand, are avoiding risk and the potential of an accident by allocating risky tasks to new employees.

There may or may not be some discretion over what task or tasks a new employee is asked to do when they start on the job. In some cases the new employee may have been certain of the tasks they would undertake from the beginning of the process of applying for the job. In other situations/jobs, such as the job of construction laborer, the new employee may be aware of the range of possible tasks, but have little certainty as to what they will actually be doing initially or on any particular day. Tasks can vary considerably in terms of safety risk, and also in their general desirability in the eyes of employees. Furthermore, it is possible that a task's safety risk and its general desirability are related, with more risky tasks also being less desirable. In fact tasks may be less desirable because they have associated safety risks.

When there is an option in terms of who performs tasks, a key question is 'who should be performing tasks with an associated safety risk component?' Without management intervention in this decision process, it is possible that a *norm* could develop within a work group or set of co-workers where the new employee is asked to complete tasks which other more senior employees find undesirable. If these undesirable tasks also have a higher safety risk, then the normative behavior of getting the new employee to do these tasks exposes the new employee to unnecessary risk. This is clearly not desirable in terms of new employee safety, as they may in fact be the least capable to undertake the tasks in safe manner.

In terms of a strategy to deal with this aspect of the adaption process, the first question relates to whether there is any discretion in the task(s) which the new employee could be asked to engage with. If there is no discretion, there is no option to minimize risk by *selective task assignment* in the initial employment period. When there is discretion, which basically means there are a number of tasks which require completing, these tasks should be ranked in terms of safety risk. Clearly, a task ranking based on objective data, such has previous accidents and incidents associated with the task, would be ideal. But in the absence of such objective data, subjective task safety rankings would be better than none at all. Subjective task safety rankings could be obtained by simply asking employees to rank the tasks in terms safety risk. Once the safety risk rank has been obtained for tasks, the key strategy is to engage new employees with tasks ranked low in terms of safety risk, and only move them on to more risky tasks once they have adapted.

7.4   Supervision

A new employee should attract the attention and guidance of their supervisor from the moment they arrive on the job. It should be part of the supervisor's job description that they provide special attention to new employees, and also that they prohibit new employees from working before supervisory guidance is provided. Chapter 4, Sect. 4.2.8 also discussed aspects of supervisor behavior. While a new employee can acquire information about job risks and safety from a number of different sources (co-workers as information sources are discussed in the next section) there may be some preference on the part of new employees to receive information from supervisors (Burns and Conchie 2014). Part of the reason driving this preference may be role related trust which a new employee has in their supervisor (Conchie and Burns 2009), whereas co-workers at least initially are generally unknown to a new employee, and there may be a degree of suspicions associated with any information they provide. There may also be considerable risk associated with helping reciprocity caused by gratitude feelings, if co-workers informally help a new employee adapt (Chap. 8, Sect. 8.6.5 discusses these issues in detail).

Two key aspects of new employee supervision are the amount of supervisory time they are given on any particular day, and the duration for which this intensive supervision can be maintained for. The amount of intensive supervision which a new employee receives on any particular day will be partly dependent on the number of other new employees starting work at the same time. When a new employee is the only new employee starting on the job, it should be possible for the supervisor to devote considerable time to their supervision. As other new employees come into the work environment the amount of supervisory time each will receive will be diluted, and as this supervision decreases the possibility that the new employee may be exposed to a safety risk or expose another employee to a safety risk will undoubtedly increase. The duration of intensive supervision will also be influenced by the frequency of new employee arrivals. If new employees are arriving on a weekly basis, a new employee may quickly find themselves with a less than adequate amount of supervision.

Supervision of new employees should be a carefully managed process, and one that takes into consideration contextual influences such as the number of new employee arrivals and their frequency of arrival. Direct supervision, clearly stated as a key role in the supervisor's job description, may be the best way to ensure a new employee behaves in a safe manner, and does not expose themselves or others to safety risks. One of the objectives of supervision should be to ensure that the new employee gains *familiarity* as quickly as possible. Goodman and Garber (1988) argued that new employees are likely to have a lack of *familiarity* with the unique characteristic of particular machinery, the specific work environment associated with the job, work methods used in the particular organization (or work group), and co-workers behavior or particular way of performing their job. Kincaid (1996) used the term '*new worker syndrome*' to explain that new employees lack familiarity with their team's procedures. A lack of familiarity is not only bad for safety (Thomas and Petrilli 2006), but also has a negative impact on productivity (Goodman and Leydon 1991). Furthermore, familiarity with co-workers appears to be central to group based transactive memory, which relates to shared group knowledge and has a positive impact on performance (Austin 2003; Lewis 2003). Thus a key aspect of supervisory behavior from both a performance and a safety perspective is helping new employees gain familiarity.

A potential point of confusion for supervisors is misinterpreting new employee experience for familiarity. For example if a new employee arrives on the job, and the supervisor ascertains that they have several years or more of experience, they may assume (incorrectly) that the new employee needs less familiarization. Familiarity is not the same as experience, and therefore the familiarity problem applies to all four types of new employee noted in Chap. 3. The primary difference between familiarity and experience is that experience relates to past circumstances, and while experience may have some ability to generalize across situations (and can play a positive role in safety Clarke et al. 2006), past circumstances and a new job are going to be different in many ways. In contrast to experience, and by definition, familiarity is specific to a particular job environment, the particular set of equipment or vehicle (see Fell et al. 1973 for safety issues associated with vehicle familiarity) used on the job, the particular set of work procedures used on the job, and the particular group of co-workers which the new employee is to work with (see Harrison et al. 2003 for a discussion of the influence of team familiarity on performance). Thus, the recruitment of an experienced worker does not, and can not, ensure familiarity.

Supervisors should also not assume that prestart training has developed a new employee's familiarity. Management organized prestart training may have provided some information and experience which will generalize to the specific situation into which the new employee has been placed and may help with familiarity development, but there will still be unique aspects which the new employee has to adapt to. For example, a prestart training program may involve instruction on how to use a particular piece of equipment that is used on the job. This will help familiarize the new employee with this aspect of their new job. However, the training equipment and the equipment used on the job will be different, even if they are exactly the same make and model, they are still two different pieces of equipment. It is not unheard of to give the senior employees the best equipment (see Chap. 4, Sect. 4.2.2), thus the equipment which the new employee maybe asked to used, while the same as they were trained on, may in fact be considerably more used, which may in fact have increased the safety risks associated with the equipment.

Supervisors have an opportunity to reduce new employee accidents by providing the intensity and duration of supervision which will ensure everyone's safety. Formalization of this responsibility in the supervisor's job description, and a clear directive from higher levels of management that this is expected, should go a long way towards reducing safety risks, and facilitating rapid new employee adaption. Clear *familiarization goals* should be set, and achieved. Supervisors can also play a role in the development of a relationship between the new employee and other co-workers. This aspect of supervision is discussed in the next section. Finally, organizations that are thinking of moving away from a traditional structure where a team or group of employees has a supervisor, to a self-directed work team structure need to appreciate the safety issues associated with this loss of supervision (see Roy 2003, for a useful discussion of other safety issues associated with the introduction of self-directed work teams).

7.5 Co-worker Compensatory Behaviors

While an organization should have formal supervisory arrangements to help ensure new employee safety, as discussed above there are factors, such as the number of new employees arriving, which may limit supervisory arrangements. To help overcome these limitations, and to ensure complete integration into a new work environment and the acquisition of familiarity, the organization and supervisors should *formally* organize input from one or more of a new employee's co-workers.

Co-workers are the individuals that 'know' the work environment, understand the idiosyncratic features of the equipment which employees may use, and appreciate the way in which individual team members, and the team collectively, works. Co-workers may in fact be better placed than supervisors to familiarize a new employee, and help ensure workplace safety (Floyde et al. 2013; Turner et al. 2012). However, co-workers may not have the resources and time to pass on their knowledge, and familiarize new employees (Delgoulet et al. 2012), and this should be provided by making the process a formal aspect of the co-workers job. Doing this also overcomes the problems that can results from new employees attempting to repay co-workers help, with their own helping acts which can and do go terribly wrong (see Chap. 8). Furthermore, co-workers are much more likely to engage with new employees and effectively share information if the process is supported by management (Nesheim and Gressgård 2014).

The general literature on teams, and their reaction to, and acceptance of new employees, suggests a complex situation which might have safety implications. This literature suggests that the degree to which team members engage with a new employee is partially determined by a concept labelled *team receptivity* (Rink et al. 2013). Team receptivity is suggested to have 3 components: *Team reflection* on exiting work processes and how a new employee might alter these and generate new ideas and ways of working; *team knowledge utilization* which reflects the team's inclination to adopt and utilize new ideas, skills, and knowledge; and *psychological/social acceptance* which reflects the team's general willingness to accept a new employee. From a safety perspective, if the a team is characterized by low reflection, low knowledge utilization and low acceptance, they are not going to be motivated to engage with the new employee for the purposes of achieving team goals (performance). However, it is unclear whether this reluctance to engage also extends to safety issues. Of course, some initial caution, and hesitation around accepting a new employee into the team may have some safety advantages. This issue is discussed further below in the section on trust development.

While the work reviewed by Rink et al. (2013) was not concerned with safety, work by Geller and colleagues has attempted to understand why co-workers will respond positively to other (including new) employees. Geller et al. (1996) investigated predictors of a concept they termed *actively caring* which is where an employee goes beyond the call of duty to ensure the safety of other employees. Their research showed that measures of personal control, group cohesion, extroversion and reactance predicted actively caring. While an organization has little control over individual difference variables such as personal control, extroversion and reactance, they can have some influence over group cohesion.

Group cohesion is strongly linked to shared beliefs and values, and intensity of normative pressures to conform (Trice and Beyer 1993). Within the safety literature considerable attention has been devoted to safety climate (e.g., Bosak et al. 2013), which also has strong links to shared beliefs and values, but specific to safety issues, and also has a normative influence on group or team member's behavior. The factors which define safety climate are reasonably well understood (e.g., Clarke 2006), and it is also clear that a strong safety climate is positively associated with safety compliance and participation (e.g., Christian et al. 2009). One of the dimensions of safety climate should be a concern for co-workers safety (Burt et al. 1998), and thus the associated safety participation dimension should be (and I say should be as there seems to be little if any research which has actually demonstrated it) co-worker safety enhancing behavior/reactions towards new employees.

Co-workers reactions to new employees have been termed compensatory behaviors (Geller et al. 1996). These are behaviors which attempt to compensate for the fact that the employee is new to the job, and needs familiarizing with all aspects of the job. Of course familiarization is required in both directions: the new employee will need to become familiar with all aspects of the job, including their co-workers behavior and attitudes, and co-workers will need to become familiar with the new employee's behavior and attitudes. The latter is particularly important for co-worker safety. New employees can be somewhat unpredictable in their behavior due to a desire to be helpful (Burt et al. 2014). Chapter 8 focuses specifically on helping behaviors which new employees may engage in and the risks these may pose to co-workers. In addition to unexpected helping behaviors, new employees can be unpredictable in how they work, and co-workers need to be very mindful of this.

Arguably the sooner that a new employee can be *familiarized*, the sooner their behavior will become predictable, and this will help ensure everyone's safety. Co-workers can help achieve new employee familiarization, but as noted above it is important that the process is formalized by management: the new employee knows that the co-worker(s) have been formally assigned to help them adapt. Formally establishing a relationship between a co-worker and a new employee, not only helps remove helping reciprocity issues, but also can deal with other limiting factors. If the interaction between an employee and a new employee is uncontrolled, it may be influenced (limited) by several factors: there will be variation in the co-workers response to new employees determined by the formal work relationship between the co-workers and the new employee, determined by the physical distance between co-workers, determined by aspects of the work (e.g., use of protective equipment) and environment (excessive noise) which either facilitate or hinder communication, and it will be determined by the time available for interactions to occur. Each of these issues is explored in some detail below.

The work relationship between employees can vary from a new employee joining a work team, to a new employee joining an organization which has a number of other employees who are indirectly working with the new employee, to a new employee working essentially on their own (although there are other employees in the organization). This variation in work relationship could be terms *co-worker distance*. In the case of a new employee joining a work team that collectively achieves work goals, co-worker distance is likely to be very small. As co-worker distance decreases, opportunity and perhaps motivation to familiarize a new employee will increase because pertinent knowledge will be available, and at the same time the risk to co-workers from the new employee will also increase. In contrast, when co-worker distance is large, other co-workers might have limited information which can help the new employee gain familiarity, they may also have limited opportunity to pass on information, and may also feel less concerned that the new employee could do something which might impact on their safety. The later point may actually decrease co-workers desire to engage with new employees.

Gaining the safety advantages that co-worker compensatory behavior can afford relies to some extent on the ability and willingness of co-workers to interact with the new employee (even when the relationship is formal). As well as the work related link between the new employee and existing employees, there will be a geographic or physical distance factor which can impact on co-worker engagement with new employees. Engagement between existing co-workers and a new employee will be easier if they work in close physical proximity to each other. Unfortunately, many high risk occupations require a degree of worker separation which precludes easy communication, yet still places each employee at risk from the behavior of other employees. The forestry operation of cutting, retrieving and loading logs is a good example. Each employee has a role to play in the operation, each can potentially be killed by a mistake made by a co-worker, yet each is separated by sufficient distance that direct verbal communication between co-workers is at best hampered, and likely to be completely restricted. In this situation it becomes very difficult for co-workers to provide compensatory behaviors and facilitate new employee familiarization. Furthermore, in this particular situation, like many others, communication isolation will be exacerbated due to the use of protective equipment such as hearing protection. An organization needs to carefully consider all of these factors when attempting to use a co-worker to help a new employee familiarize.

A further factor which can influence co-worker compensatory behaviors is the performance demands which are placed on them. Interaction with a new employee requires time, time which will be taken away from other productive activities. A number of studies have shown a relationship between workload/performance pressure and safety (e.g., see Christian et al. 2009 for a review), with safety decreasing as performance demands increase. One mechanism which may explain these findings is the limitations which high performance demands place on employees, and in particular how they limit employees' ability to provide compensatory behaviors for new employees. Arguably new employee adaption, and their familiarization, will be slower in a situation where performance demands placed on co-workers preclude effective interaction with new employees. Thus any formal arrangement between a co-worker and a new employee needs to allow the co-worker the time necessary to provide the degree of support the new employee needs.

Co-worker compensatory behaviors may also be determined by how well the co-worker knows the new employee (Burt et al. 2008). The acquisition of knowledge about a new employee is at the basis of the finding that similarity between a new employee and co-workers is likely to prompt acceptance (e.g., Joardar 2007; Ziller and Behringer 1960). Basically, if the new employee appears to share similar values, interests and attitudes to exiting co-workers—he or she is likely to be liked and responded to in a positive way. This result is consistent with findings from basic social psychology research which have demonstrated that individuals tend not

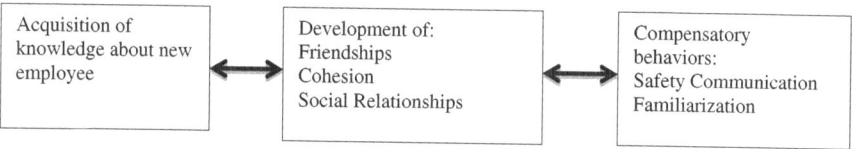

Fig. 7.1 Interaction between knowledge of new employees and the provision of compensatory behaviors

to want to help strangers (e.g., *the bystander apathy effect*; see Garcia et al. 2002), but are likely to help when the 'other' is a friend (e.g., Bell et al. 1995).

Burt et al. (2008) developed a model of the development of *considerate and responsible co-worker behavior*: a model of the determinants of safety related compensatory behaviors. Figure 7.1 shows the main features of the model. In one study Burt et al. (2008) sampled workers in the forestry and construction industries, and found that *caring* about co-workers (measured by the Considerate and Responsible Employee (CARE) scale, Burt et al. 1998—Chap. 9) was positively correlated with the amount of knowledge an employee had about their co-workers. Based on these findings, any process which allows co-worker (team members) to get to know new employees (helps to build the social fabric of a work team) is likely to be associated with an increase in compensatory behaviors towards new employees, quicker new employee adaptation, and faster development of familiarity. Of course, people don't like everyone they meet, and there may be situations where the knowledge gained about a new employee is such that a co-worker is not interested in helping them adjust (although they may still do so partly in order to ensure their own safety).

From the points made in this section a strategy to ensure that co-workers provide compensatory behaviors towards new employees needs to realize the limitations which co-worker distance, performance demands and occupational/job characteristics (e.g., physical distances, protective equipment) will place on the process. The strategy also needs to recognize the advantages which can be achieved by management formalizing and supporting interactions between co-workers and new employees, and by socialization processes which allows existing employees to get to know new employees as quickly as possible. Where limitations are apparent it will be vital that steps are taken by management to ensure co-workers can provide the necessary compensatory behaviors.

7.6 Changes in New Employee Behavior

One of the most obvious differences in the behavior of a new employee over the initial period of employment is their information seeking behavior. In order to deal with the uncertainty and lack of familiarity associated with entry into a new job, a new employee is likely to engage in a range of information seeking behaviors. As uncertainty is reduced and familiarity is gained by the acquisition of information, the new employee's information seeking behaviors will decline. Miller and Jablin (1991) developed a model of newcomer information seeking tactics. Their model describes 7 information seeking tactics: *Overt* and *indirect* questioning of co-workers and/or supervisors, questioning *third parties*, *testing limits* where, for example, the new employee tries a behavior and waits for others to respond, *disguising conversations*, where information is sort but not directly asked for, *unobtrusively observing* others specific behavior in order to gain a specific understanding, and *surveillance* which is a more general unfocused monitoring of others behavior. It is clear that a number of these tactics, if not all of them, could have associated safety risks. At the very least, they all will take the new employees, and in some cases both the new employee and other workers, focus off the job, and we know that situational awareness is a key factor in the maintenance of safety.

Adaption and familiarization will also be paralleled by other changes in the new employee's behavior. Some of these changes will be the result of learning how tasks are performed, and some may be adjustments, or changes, to the way the new employee has performed in the past. The latter is more likely when a new employee with previous work experience arrives and discovers that a job they have performed previously in another organization, with a different group of co-workers, is performed differently in the new situation. The new employee essentially has to unlearn task behaviors, and adjust or adapt to how the tasks are performed in the new situation.

Exactly how a new employee adapts their behavior to a new job was the focus of Nicholson's (1984) role transition theory. While not writing about safety, Nicholson suggested that a new employee's adaption could be classified into 1 of four outcomes: *replication, absorption, determination* and *exploration*. Replication is characterized by minimal adjustment on the part of the new employee to the way they have worked in the past, and no attempt on their part to adjust or change the job they have entered. Absorption sees significant adjustment on the part of the new employee and changes in the way they complete the job's task, and no attempt to impose changes on the job or organization. *Determination*, in contrast, is characterized by the new employee changing the way the job is performed to suit their way of work. Finally, *exploration* involved a degree of change from the new employee and also a degree of change to the way the job is performed. Clearly, from a co-worker (and organization) perspective replication is likely to cause the least disruption and allow for the greatest predictability in the new employee's behavior. On the other hand, determination could be very disruptive and undermine co-workers ability to predict how the new employee is going to behave.

While it is clear that a new employee's behavior will (more or less) go through a period of adjustment in their initial employment period, it is also likely that their verbal behavior will change (also see Chap. 4. sect. 4.2.9). It is well established that employees' voicing of safety concerns is an important aspect of workplace safety (Hofmann and Morgeson 1999; Kath et al. 2010). To a considerable extent the development of voicing behaviors will be dependent on the development of trust relationships with supervisors (Conchie et al. 2012), and with co-workers

(Tucker et al. 2008). Unfortunately, (or perhaps fortunately, see Sect. 7.7) trust relationship take time to develop. Thus, new employees may initially be particularly reluctant to speak up about safety concerns, until they determine the type of reaction they are likely to get from co-workers, supervisors and management. New employees may refrain from voicing safety concerns from a fear of retaliation (Collinson 1999; Jeffcott et al. 2006), a fear of being viewed negatively (e.g., Milliken et al. 2003), or an assumption that safety violations may be normalized behavior (Ashford and Anand 2003).

The final type of behavior which might be particularly associated with new employees are random acts of helping. Burt et al. (2014) found evidence that helping declines as job tenure increases. While society generally looks positively on acts of helping, and there is a vast literature on the value of organizational citizenship behaviors, helping acts can be unpredictable, and can put both the new employee and co-workers at risk. Chapter 8 is devoted to a discussion of helping behaviors, motivations behind helping and the associated safety risks. For now it is important to realize that new employees may be particularly likely to try and be helpful. Such behavior may result in the new employee being in unpredictable places and doing unpredictable things. Even though their intentions are to be helpful, the consequences may disastrous.

Every organization should have a strategy to deal with changes in new employee behavior. The strategy should focuse on making co-workers aware of how new employees are likely to behave and how this may change over time. Co-workers need to be encouraged to be accepting and supporting of new employee information seeking behaviors. Co-workers need to expect that new employee behavior may be different from previous incumbents in the job, that they may engage in random acts of helping, and that new employees may not perform the job in the way that co-workers are expecting. Furthermore, new employees may be reluctant to voice safety concerns, as they will not initially know how co-workers (and management) will react to such voicing. In the past a job incumbent may have immediately informed co-workers when a hazard becomes apparent, whereas a new employee may not initially do this.

7.7 Trust Development

Trust plays a central role in safety (Conchie et al. 2006). Studies have shown links between positive safety outcomes, and trust in management (e.g., DePasquale and Geller 1999; Kath et al. 2010; Luria 2010), and trust in co-workers (e.g., Tharaldsen et al. 2010). Trust is also a key aspect of a positive safety culture (Burns et al. 2006), influences safety attitudes (Walker 2013), and influences the effectiveness of risk communication (Conchie and Burns 2008; Twyman et al. 2008). While there are clear safety benefits associated with trust, safety benefits can also come from distrust (Conchie and Donald 2008), and this is likely to particularly be the case in relation to new employees in their initial period of employment. Trust can reduce an employees' inclination to monitor and safeguard against risks from a new employee, and can decrease their judgment of new employees based on *their* behaviors (Conchie and Donald 2008; McEvily et al. 2003). Thus encouraging team members not to initially trust new employees may help increase behaviors aimed at ensuring the safety of new employees, and indeed themselves.

A key proposition of the *social capital theory* of turnover costs is that turnover reduces the level of collective goal focus and shared trust (Leana and van Buren 1999). Thus a vacancy which has prompted the employment of a new employee will have had a negative impact on trust within the workplace. The same argument can be applied when a new employee is acquired due to an increased need for human capital. Therefore there is always going to be a trust development process associated with the arrival of a new employee. The new employee will develop more or less trust in management, supervisors and co-workers, and management, supervisors and co-workers will develop more or less trust in the new employee (Jeffcott et al. 2006). A new employee will also develop more or less trust in the systems and processes of the organization, as well as in the equipment they are asked to use to complete their work.

Supervisors' interactions with new employees, what might be terms their leadership style, will play an important role in the establishment of trust-relationships. Evidence is mounting that a transformational leadership style, where leaders develop affective bonds with their employees will help facilitate trust development and positively influence safety (Conchie 2013), as well as positively influence performance outcomes (Schaubroeck et al. 2011). Supervisors should of course develop a safety-specific trust relationship with a new employee based on evidence from their behavior, not based on assumptions. While supervisors may be somewhat insulated from the adverse impact of new employee's behavior, they should consider new employees as potential sources of safety risk until proven otherwise. Co-workers are likely to be the most vulnerable in terms of the impact of unsafe behavior from new employees. Thus from the perspective of co-workers, it is advisable to be careful and ensure that any trust which is given to a new employee is deserved.

From the general literature on trust we know that there are 3 key elements associated with trust development: ability, benevolence and integrity (Mayer et al. 1995). If we consider these in relation to a new employee and safety, ability encompasses the knowledge, and skills which the new employee brings to the job, and also their effective use of these. Benevolence is essentially a positive orientation, or in relation to safety a positive commitment to the safety of self and others. Integrity refers to adherence to safety rules, safety compliance and participation. Each factors takes time to be shown and witnessed by others, and as such trust development should take time and be based on evidence of a new employee's safety related ability, benevolence and integrity. Another way of thinking of this is that there are 4 key dimensions of trust: *commitment* (e.g., the new employee is committed to performing their job in a safety manner, *competence* (e.g., the new employee has the knowledge, skills, abilities to performs their job safely), *caring* (e.g., the new employee cares how their behavior or lack of behavior will impact on others safety), and *predictability* (e.g., the new employee is consistent in how they behavior).

There are two key variables associated with trust and new employee's entry into a workplace: *risk potential* and *time*. Clearly risk potential decreases with time (job tenure). Trust takes time to develop, and in terms of maintaining workplace safety, trust should not be given lightly. An organization can simply let trust develop, or it can attempt to manage its development. Certainly, all organizations should provide for trust to develop by employing new employees that show evidence of *commitment, competence* and *caring. Predictability* on the other hand needs to be demonstrated. In the next section I suggest a strategy by which predictability can be used as the foundation for trust development.

7.7.1 A Trust Building Strategy

Organizations should consider adopting strategies which shift perceptions of new employees from distrusted to trusted based on evidence of a decrease in potential risk within a reasonable time period, and thus gain the safety advantage that both distrust and trust can bring. Burt and Hislop (2013) (also see Burt and Stevenson 2009) suggested that a process where all new employees wear a specific color safety vest during their initial period of employment could have significant positive safety outcomes. The 'new employee vest' would immediately identify the new employee as a potential safety risk (signal that their predictability in terms of safety is yet to be established), and warn co-workers to be especially cautious about their safety. Organizations could also adopt a process of moving a new employee (their vest color) to a standard team member vest (color) based on the new employee achieving a set level of safety specific trust rating from co-workers. Thus it would be the new employee's co-workers that decide when they are ready to trust the new employee.

The process could involve a new employee's co-workers rating their safety behavior each week. Essentially the objective is determine that the new employee does behavior in a safety way, is predictable in how they behave, and based on this can be trusted to ensure everyone's safety to the best of their ability. Once a predetermined level of safety rating and consistency of safety rating is achieved, it is signaled by the allocation of a team vest that the team accepts the new employee as a safe co-worker. The process may have several advantages: it signals that there are safety risks associated with new employees, that the organization accepts these risks exist, that new employees have to earn the trust of their co-workers, and it actively engages employees in the development of safety-specific trust in new employees.

Table 7.1 Key issues to consider during a new employees initial period of employment

Key issue	Management response
Is there discretion over task assignment?	Assign tasks based on safety risk
Supervisor attention and guidance for new employees	New employee focus is a key role in the supervisor's job description
	Set familiarization goals
	Adjust supervision to accommodate frequency of new employee arrival
Coworker support for new employee adaption	Formalize the link between co-workers and new employee
	Control helping reciprocity
	Recognize and manage contextual limitations that can restrict co-worker compensatory behaviors
Understand new employee initial behavior	Manage new employee behavior and alert co-workers to: Information seeking behavior Desire to change role/tasks Employee silence/voicing Random acts of helping
Understand trust development	Management trust development to gain advantage of both distrust and trust

7.8 Conclusions

This chapter has discussed a number of factors associated with a new employees initial period of employment. Arguably the first 3 months of employment is characterized by the greatest level of safety risk. This is not to say that after 3 months or so a new employee is not going to have an accident, clearly there will risks that go well beyond the first 3 months. However, the first 3 months of work in a new job is a period of time during which the organization should attempt to manage a number of processes, and careful consideration of each should a go a long way towards reducing new employee accidents. Table 7.1 shows the key points made in this chapter, and highlights the key management response.

References

Ashford, B. E., & Anand, V. (2003). The normalization of corruption in organizations. *Research in Organizational Behavior, 25,* 1–52.

Ashford, S. J., & Black, J. S. (1996). Proactively during organizational entry: The role of desire for control. *Journal of Applied Psychology, 81,* 199–214.

Austin, J. R. (2003). Transactive memory in organizational groups: The effects of content, consensus, specialization, and accuracy on group performance. *Journal of Applied Psychology, 88*(5), 866–878.

Bauer, T. N., & Green, S. G. (1994). Effect of newcomer involvement in work-related activities: A longitudinal study of socialization. *Journal of Applied Psychology, 79,* 219–223.

Bell, J., Grekul, J., Lamba, N., & Minas, C. (1995). The impact of cost on student helping behaviour. *Journal of Social Psychology, 135*(1), 49–56.

Bosak, J., Coetsee, W. J., & Cullinane, S. (2013). Safety climate dimensions as predictors for risk behavior. *Accident Analysis and Prevention, 55*, 256–264.

Burns, C., & Conchie, S. (2014). Risk information source preferences in construction workers. *Employee Relations, 36*(1), 70–81.

Burns, C., Mearns, K., & McGeorge, P. (2006). Explicit and implicit trust within safety culture. *Risk Analysis, 26*(5), 1139–1150.

Burt, C. D. B., Banks, M., & Williams, S. (2014). The safety risks associated with helping others. *Safety Science, 62*, 136–144.

Burt, C. D. B., Gladstone, K. L., & Grieve, K. R. (1998). Development of the considerate and responsible employee (CARE) scale. *Work and Stress, 12*(4), 362–369.

Burt, C. D. B., & Hislop, H. (2013). Developing safety specific trust in new recruits: The dilemma and a possible solution. *Journal of Health, Safety and Environment, 29*(3), 161–173.

Burt, C. D. B., Sepie, B., & McFadden, G. (2008). The development of a considerate and responsible safety attitude in work teams. *Safety Science, 46*, 79–91.

Burt, C. D. B., & Stevenson, R. J. (2009). The relationship between recruitment processes, familiarity, trust, perceived risk and safety. *Journal of Safety Research, 40*, 365–369.

Chan, D., & Schmitt, N. (2000). Interindividual differences in intraindividual changes in proactivity during organizational entry: A latent growth modelling approach to understanding newcomer adaptation. *Journal of Applied Psychology, 85*, 190–210.

Christian, M. S., Bradley, J. C., Wallace, J. C., & Burke, M. J. (2009). Workplace safety: A meta-analysis of the roles of person and situation factors. *Journal of Applied Psychology, 94*(5), 1103–1127.

Clarke, S. (2006). The relationship between safety climate and safety performance: A meta-analytic review. *Journal of Occupational Health Psychology, 11*(4), 315–327.

Clarke, D. D., Ward, P., Bartle, C., & Truman, W. (2006). Young driver accidents in the UK: The influence of age, experience, and time of day. *Accident Analysis and Prevention, 38*(5), 871–878.

Collinson, D. L. (1999). Surviving the rigs: Safety and surveillance on North Sea oil installations. *Organizational Studies, 20*, 579–600.

Conchie, S. M. (2013). Transformation leadership, intrinsic motivation, and trust: A moderated-mediated model of workplace safety. *Journal of Occupational Health Psychology, 18*(2), 198–210.

Conchie, S. M., & Burns, C. (2008). Trust and risk communication in high-risk organizations: A test of principles from social risk research. *Risk Analysis, 28*(1), 141–149.

Conchie, S. M., & Burns, C. (2009). Improving occupational safety: Using a trusted information source to communicate about risk. *Journal of Risk Research, 12*(1), 13–25.

Conchie, S. M., & Donald, I. J. (2008). The functions and development of safety specific trust and distrust. *Safety Science, 46*, 92–103.

Conchie, S. M., Donald, I. J., & Taylor, P. J. (2006). Trust: Missing piece(s) in the safety puzzle. *Risk Analysis, 26*(5), 1097–1104.

Conchie, S. M., Taylor, P. J., & Donald, I. J. (2012). Promoting safety voice with safety-specific transformational leadership: The mediating role of two dimensions of trust. *Journal of Occupational Health Psychology, 17*(1), 105–115.

Delgoulet, C., Gaudart, C., & Chassaing, K. (2012). Entering the workforce and on-the-job skills acquisition in the construction sector. *Work, 41*, 155–164.

DePasquale, J. P., & Geller, E. S. (1999). Critical success factors for behavior-based safety: A study of twenty industry wide applications. *Journal of Safety Research, 30*(4), 237–249.

Endsley, M. (1995). Towards a theory of situation awareness in dynamic systems. *Human Factors, 37*, 32–64.

Fell, J. C., Mudrowsky, E. F., & Tharp, K. J. (1973). A study of driver experience and vehicle familiarity in accidents. *Accident Analysis and Prevention, 5*, 261–265.

Floyde, A., Lawson, G., Shalloe, S., Eastgate, R., & D'Cruz, M. (2013). The design and implementation of knowledge management systems and e-learning for improved occupational health and safety in small to medium sized enterprises. *Safety Science, 60,* 69–76.

Garcia, S. M., Weaver, K., Moskowitz, G. B., & Darley, J. M. (2002). Crowed minds: The implicit bystander effect. *Journal of Personality and Social Psychology, 83*(4), 843–853.

Geller, E. S., Roberts, D. S., & Gilmore, M. R. (1996). Predicting propensity to actively care for occupational safety. *Journal of Safety Research, 27*(1), 1–8.

Goodman, P. S., & Garber, S. (1988). Absenteeism and accidents in a dangerous environment: Empirical analysis of underground coal mines. *Journal of Applied Psychology, 73,* 81–86.

Goodman, P. S., & Leyden, D. P. (1991). Familiarity and group productivity. *Journal of Applied Psychology, 76*(4), 578–586.

Harrison, D. A., Mohammed, S., McGrath, J. E., Florey, A., & Vanderstoep, S. W. (2003). Time matters in team performance: Effects of member familiarity, entrainment, and task discontinuity on speed and quality. *Personnel Psychology, 56,* 633–669.

Hofmann, D. A., & Morgeson, F. P. (1999). Safety-related behavior as a social exchange: The role of perceived organizational support and leader-ember exchange. *Journal of Applied Psychology, 84,* 286–296.

Jeffcott, S., Pidgeon, N., Weyman, A., & Walls, J. (2006). Risk, trust and safety culture in U.K. train operating companies. *Risk Analysis, 26*(5), 1105–1121.

Joardar, A., Kostova, T., & Ravlin, E. C. (2007). An experimental study of the acceptance of a foreign newcomer into a workgroup. *Journal of International Management, 13,* 513–537.

Kath, L. M., Magley, V. J., & Marmet, M. (2010a). The role of organizational trust in safety climate's influence on organizational outcomes. *Accident Analysis and Prevention, 42,* 1488–1497.

Kath, L. M., Marks, K. M., & Ranney, J. (2010b). Safety climate dimensions, leader-member exchange, and organizational support as predictors of upward safety communication in a sample of rail industry workers. *Safety Science, 48,* 643–650.

Kincaid, W. H. (1996). Safety in the high-turnover environment. *Occupational Health and Safety, 65,* 22–25.

Leana, C. R., & Van Buren, H. J. I. I. I. (1999). Organizational social capital and employment practices. *Academy of Management Review, 24,* 538–555.

Lewis, K. (2003). Measuring transactive memory systems in the field: Scale development and validation. *Journal of Applied Psychology, 88*(4), 587–604.

Luria, G. (2010). The social aspects of safety management: Trust and safety climate. *Accident Analysis and Prevention, 42,* 1288–1295.

Mayer, R. C., Davis, J. H., & Schoorman, F. D. (1995). An integrative model of organizational trust. *The Academy of Management Review, 20*(3), 709–734.

McEvily, B., Perrone, V., & Zaheer, A. (2003). Trust as an organizing principle. *Organization Science, 14,* 91–103.

Miller, V. D., & Jablin, F. M. (1991). Information seeking during organizational entry: Influences, tactics, and a model of the process. *Academy of Management Review, 16*(1), 92–120.

Milliken, F. J., Morrison, E. W., & Hewlin, P. F. (2003). An exploratory study of employee silence: Issues that employees don't communicate upward and why. *Journal of Management Studies, 40*(6), 1453–1476.

Nesheim, T., & Gressgård, L. J. (2014). Knowledge sharing in a complex organization; Antecedents and safety effects. *Safety Science, 62,* 28–36.

Nicholson, N. (1984). A theory of work role transitions. *Administrative Science Quarterly, 29,* 172–191.

Rink, F., Kane, A., Ellemers, N., & Van der Vegt, G. (2013). Team receptivity to newcomers: Five decades of evidence and future research themes. *The Academy of Management Annals, 7*(1), 247–293.

Roy, M. (2003). Self-directed workteams and safety: A winning combination? *Safety Science, 41,* 359–376.

Schaubroeck, J., Lam, S. S. K., & Peng, A. C. (2011). Cognition-based and affect-based trust as mediators of leader behavior influences on team performance. *Journal of Applied Psychology, 96*(4), 863–871.

Tharaldsen, J. E., Mearns, K. J., & Knudsen, K. (2010). Perspectives on safety: The impact of group membership, work factors and trust on safety performance in UK and Norwegian drilling company employees. *Safety Science, 48*, 1062–1072.

Thomas, M. J. W., & Petrilli, R. (2006). Crew familiarity: Operational experience, non-technical performance, and error management. *Aviation, Space and Environmental Medicine, 77*(1), 41–45.

Trice, H. M., & Beyer, J. M. (1993). *The cultures of work organizations.* Englewood Cliffs, NJ: Prentice-hall.

Tucker, S., Chmiel, N., Turner, N., Hershcovis, S., & Stride, C. B. (2008). Perceived organizational support for safety and employee safety voice: The mediating role of coworker support for safety. *Journal of Occupational Health Psychology, 13*(4), 319–330.

Turner, N., Stride, C. B., Carter, A. J., McCaughey, D., & Carroll, A. E. (2012). Job demands-control-support model and employee safety performance. *Accident Analysis and Prevention, 45*, 811–817.

Twyman, M., Harvey, N., & Harries, C. (2008). Trust in motives, trust in competence: Separate factors determining the effectiveness of risk communication. *Judgment and Decision Making, 3*(1), 111–120.

Walker, A. (2013). Outcomes associated with breach and fulfillment of the psychological contract of safety. *Journal of Safety Research, 47*, 31–37.

Ziller, R. C., & Behringer, R. D. (1960). Assimilation of the knowledgeable newcomer under conditions of group success and failure. *The Journal of Abnormal and Social Psychology, 60*, 288–291.

Chapter 8
New Employee Helping Behaviors

8.1 Introduction

Figure 1.1 shown in Chap. 1 has a box labeled new employee helping. This chapter describes how helping can be good in some work situations where it might be described as organizational citizenship behavior, and yet in other situations helping can be very dangerous. Evidence is presented showing that helping decreases as organizational tenure increases, or in other words, new employees appear to engage in more helping behaviors compared to more senior employees. This makes the safety risks associated with helping a particular problem associated with new employees. A number of different safety risks which can results from helping are discussed. Finally, the chapter explores a number of reasons why new employees may engage in helping and makes a number of suggestions of how this safety risk associated with new employees can be managed.

8.2 Helping

Being helped is generally a positive and pleasant experience. However, helping can on occasions go terribly wrong. In the 1980s, I witnessed the death of a worker at a rubbish processing plant. The plant accepted rubbish delivered by the public (which is what I was doing) and also rubbish delivered by commercial rubbish collection vehicles. On the day of the accident, I was unloading my rubbish into a large concrete pit. A commercial rubbish collection vehicle (truck) was reversing into the pit in order to eject its load. This vehicle stopped momentarily, and one of the workers jumped out and ran behind the truck and proceeded to direct the driver as he reversed the vehicle. The worker on the ground then ran toward the reversing vehicle and attempted to jump onto the tail board of the truck, but slipped and was run over by the rubbish truck and killed. I was called as a witness at the coroner's

© Springer International Publishing Switzerland 2015
C.D.B. Burt, *New Employee Safety*,
DOI 10.1007/978-3-319-18684-9_8

court. The courts conclusion, after evidence from all parties was presented, was that the deceased employee was attempting to help speed up the unloading process by releasing the rear door locks before the vehicle had come to a stop. His attempt to help had cost him his life.

It is not hard to find other examples of good intentioned behavior, what society typically refers to as helping or pro-social behavior, leading to the injury of the helper or to the injury of the person being helped. For example, Organ et al. (2006) provide the following example: 'Sam did, of course, earn Dennis's considerable gratitude, a token of which was demonstrated later in the evening when Dennis helped Sam, in the course of which Dennis toppled a 200-pound roll of paper onto his own foot and broke his toe.' (p. 3). Within the organizational psychology literature, the terms *organizational citizenship behavior* and/or *contextual performance* (Borman and Motowildo 1997) are often used to describe helping behaviors which are not formally expected from an employee, yet contribute to effective functioning (Organ, 1988; Podsakoff et al. 2000). It is important to note that helping is but one type of organizational citizenship behavior (see Carpenter et al. 2014; Chiaburu et al. 2011; Hoffman et al. 2007; Jiao et al. 2013; Nielsen et al. 2009; Organ and Ryan 1995; Spector and Che 2014; Whitman et al. 2010 for meta-analytic reviews of the organizational citizenship behavior literature). The specific act of helping has also received a reasonable amount of research attention (e.g., De Jong et al. 2007; Dyne and LePine 1998; Naumann and Ehrhart 2011; Oosterhof et al. 2009; Van der Vegt et al. 2006; Venkataramani and Dalal 2007), as has safety-specific helping (e.g., Didla et al. 2009; Hofmann et al. 2003; Gyekye and Salminen 2005).

The literature has generally been broadly encouraging of employees 'helping.' Indeed, in many work environments helping may provide considerable advantage, by increasing organizational effectiveness (e.g., Podsakoff and Mackenzie 1994) and productivity (e.g., Gyekye and Salminen 2005), and enhancing the quality of service to customers (e.g., Bell and Menguc 2002). Furthermore, safety-specific helping should improve safety. It is therefore appropriate for organizational citizenship behaviors or contextual performance to be encouraged. However, where the work which is being undertaken, or the environment in which a job is being performed, has associated safety risks, the helping aspect of organizational citizenship behavior might lead to an accident or injury and needs to be carefully managed (Burt et al. 2014).

8.3 New Employees: Helping and Job Tenure

New employees are perhaps the least prepared in terms of the risk factors associated with helping (discussed below), yet new employees seem to be more likely to actually try and help. For example, new employees may not fully understand task risks, may not have fully developed the abilities and skills (or understand the abilities and skills) required for the task they are wanting to help with, and may not realize the importance of acknowledging their interest in helping. Burt et al. (2014)

Table 8.1 Descriptive statistics for participant tenure, and correlations between helping outcome measures and tenure reported by Burt et al. (2014)

| Study | Tenure in months | | | Helped and increased safety risk for yourself | Helped and increased safety risk for the person you helped | Helped and increased safety risk for another employee | Overall helping safety risk measure |
	Mean	SD	Min and Max				
Study 1	163.6	139.0	1–552	−0.10	−0.12*	−0.15***	−0.14***
Study 2	85.0	119.6	1–540	−0.18	−0.23***	−0.21**	−0.23***

*p = 0.06; **p = 0.053; ***p < 0.05

conducted a study on the relationship between *helping* and workplace safety and examined the relationship between helping and employee tenure. Two studies identified a number of significant negative relationships between tenure and helping which had resulted in a safety risk: Study 1 sampled 222 participants from two processing factories and a road maintenance/construction company in New Zealand, while Study 2 sampled 79 participants working in high-risk jobs from the construction, engineering, electrical, road construction, shipping, and healthcare industries in New Zealand.

Participants in the studies were asked to indicate the frequency of occurrences when their helping had resulted in a safety risk for them, for the person they were helping or for another employee. An overall composite measure was also generated by combining the responses to the latter three questions. Some key results from Burt et al. (2014) are shown in Table 8.1. The negative correlations shown in Table 8.1 suggest that new employees (those with less tenure) engage in more helping behaviors which were perceived as resulting in a safety risk. These findings are also consistent with research which indicates that humans can learn from their mistakes (e.g., Tjosvold et al. 2004). That is, more senior workers may have learnt that helping in some circumstances can be risky, and as such, refrain from such helping. In contrast, somewhat enthusiastic new employees may not have learnt this, and thus engage more frequently in helping that results in safety risks. The next section examines three key factors associated with helping which make it risky and which may also help explain why new employees report more occurrences of helping which results in safety issues.

8.4 Why Is Helping Risky?

There are several key determinants of whether helping might have safety risks and result in an accident/incident. First is a consideration of the risks associated with the task. For example, the two tasks of helping a colleague to supervise a test, and helping a colleague chain down a load of logs on a truck, carry very different safety risks. The latter example implies there is a *continuum of associated safety risk* onto which all tasks can be placed, and engaging in helping with risky tasks (or with a task being performed in a risky environment) is not advisable. Arguably, a new employee may have little (or less) knowledge of the hazards and safety risks associated with the task(s) they are attempting to help with, and this makes the *new employee helper* vulnerable. The literature on organizational citizenship behavior has tended to ignore the inherent safety risk associated with the task which the employee is helping with. While it is appropriate to encourage helping with tasks which have no associated safety hazards or risks, and to encourage safety-specific helping, it is not appropriate for employees to help with tasks which have associated safety issues.

The second factor to consider is the ability of the helper to actually provide the help. An employee may feel confident that they have the necessary knowledge, skills, and abilities required to help, yet they may (while having the best of intentions) be entering into a situation which they are not equipped to deal with. Chapter 7 provides an extensive discussion of the adaption processes which new employees go through. The issue of familiarization is also discussed in Chap. 7. The need for adaption and familiarization, along with the limited ability of past experience to generalize to a new situation/job (see Chap. 3), means that a new employee is always going to have less ability to help safely compared to a senior employee that has adapted to the work place and is familiar with all aspects of it.

The third factor is the interaction between the helper and person being helped. This interaction will vary along a continuum which ranges from a *verbal offer of help* (e.g., 'I can help,' 'Would you like help,' 'Can I help') to *unannounced and unacknowledged helping*. In *unannounced helping*, the employee steps into help but does not tell (or ask) any other employee first, and in *unacknowledged helping*, the helper provides help but does not tell any other employee what they have done after the fact. Together these two components of the interaction factor are referred to as *acknowledgment* in the following discussion. In the verbal offer situation, the helper may be rejected, and the employee being helped will know the helper is there —both mechanisms should help protect the helper. In contrast, *unannounced and unacknowledged helping* is characterized by the employee that is being helped not knowing that the help is being given or has been given. This situation is extremely dangerous, as it can result in an employee being in a location where other employees are not expecting them to be, or aspects of the work environment being changed by the helper and thus different to that which an employee is expecting. Consider the example of a new employee helping a co-worker chaining down a load of logs on a truck. Part of this operation can involve chains being thrown over the load and secured. A new employee, in order to help by securing one of the chains, that moves '*unannounced*' into the area where a chain is landing may well be injured for their efforts.

Communication is particularly important in helping situations. Helping is a deviation from normal work behavior, and by definition is not expected behavior. Of course, a work group may develop a norm where helping is expected (see Naumann and Ehrhart 2011). Yet, individual instances of helping will still be hard to predict. Arguably, if an employee is going to do something which is outside their job boundaries, they need to let other employees know. An employee who is unexpectedly in someone else's work space or who is unexpectedly engaging in a behavior which other employees were not expecting or aware of (despite their good intentions), may well expose themselves to danger. Furthermore, if help is provided for an employee and they are not made aware of this, that employee may be exposed to a safety issue simply because they do not know what has been done. Employees need to be particularly careful of this scenario when it occurs in association with shift changes. There is growing evidence of the safety issues associated with shift changes (e.g., Matric et al. 2010; Mayor et al. 2011; Jiag et al. 2002), although the issue of helping and shift changes does not appear to have been researched.

Figure 8.1 shows a representation of the three factors (task safety risk, helper ability, and acknowledgment) associated with helping and the degree of risk these factors can *combine* to produce. In order to help clarify the issues, the three factors have each been scaled from 1 to 7 such that a lower scale score equals less helping-related risk. Thus for *task risk,* low task risk = 1 and high task risk = 7, for helper ability low ability = 7 and high ability = 1, and for acknowledgment none = 7 and complete = 1 (complete acknowledgment is where the helper announces the offer of help ('Can I help'), the help is accepted ('Yes'), and thus the person being helped knows what the helper is doing or about to do). Overall, there is less risk associated with helping when task risk is low, helper ability is high, and acknowledgment is

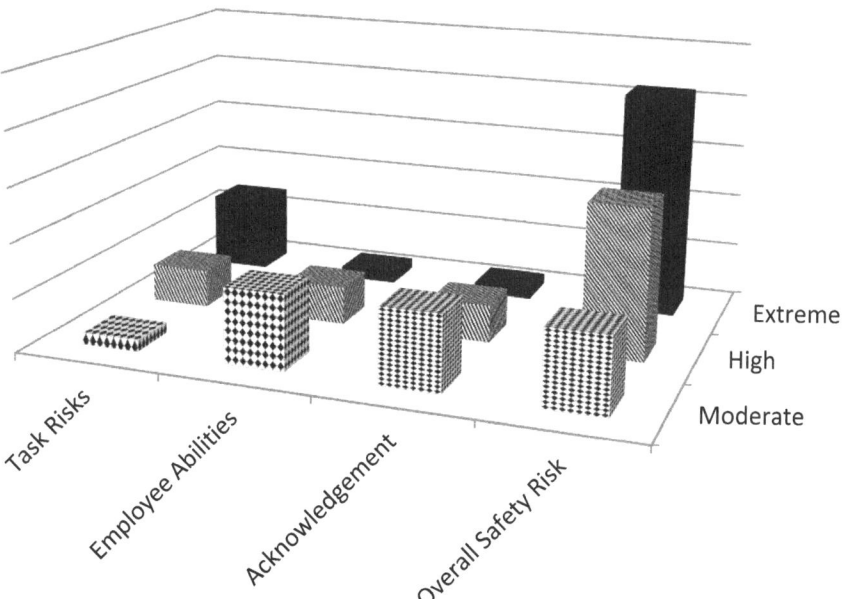

Fig. 8.1 A model of the relationship between task risks, employee abilities, helping acknowledgment, and the overall safety risk associated with helping

complete. Inspecting the rows of bars in Fig. 8.1, the solid black row of bars is labeled as having an extreme overall safety risk, as task risk is high, helper ability is low, and there is no acknowledgment. In contrast, the middle row of bars has a high overall safety risk and is characterized by moderate task risk, moderate helper ability, and incomplete acknowledgment. Finally, the row of bars with the diamond pattern has a moderate overall safety risk and is characterized by low task risk, helper ability, and acknowledgment. The reason that the row of bars with the diamond pattern produces a moderate overall safety risk classification is, for example, because the act of leaving ones job to help another employee can create a situation where something happens in the helpers job (in their absence) which can lead to an accident. This possibility, along with other possible scenarios of how helping can be risky, is discussed further below.

8.5 Other Risk Mechanisms Associated with Helping

Figure 8.1 shows how task risk, helper ability, and acknowledgment can combine to produce a varying level of safety risk associated with helping. Furthermore, the examples given above suggest how a helper can be injured or killed from the act of helping. In addition to these factors, several other factors associated with helping can result in safety risks and indeed accidents. These are referred to as the *forgetting*, the *absence,* the *hazard creation,* and the *time pressure* mechanisms. Each mechanism is discussed below. Burt et al. (2014) found evidence supporting each of these mechanisms.

8.5.1 The Forgetting Mechanism

Leaving one's job to help another employee could be described as a self-initiated interruption. Another view of this is that helping another employee takes the helpers attention away from their job. Not surprisingly, there is research evidence showing that inattention, interruptions, and distractions have negative workplace safety outcomes (e.g., Cohen et al. 1985; Tremblay et al. 2012). The disruption which helping can create has the potential to disrupt cognitive aspects of an employee's tasks. For example, the employee's ability to remember what they had just done or what they were about to do. Burt et al. (2014) found evidence of 3 types of forgetting associated with helping: the employee forgets at what point they were at in their work; forgets something which they had planned to do; and forgets what they had just done prior to leaving their job to help another. Forgetting as a result of distraction from helping another could also be classified as a cognitive failure, and cognitive failures have been linked to negative safety outcomes (e.g., Norman 1981; Wallace and Chen 2005; Wallace and Vodanovich 2003).

8.5.2 The Absence Mechanism

The departure from one's job to help another employee could result in something unexpected happening in the helper's job (in their absence) which would not have happened had they remained focused on their work. Many jobs require an employee to monitor system components (e.g., pressure, temperature, and items moving on a conveyor) and to make system adjustments in response to changes. A failure to monitor systems associated with a job as a result of an absence associated with helping another employee could result in a system entering a state which could lead to an accident.

8.5.3 The Hazard Creation Mechanism

Helping can involve the helper changing aspects of a workplace, and in doing this the helper can create a hazard. For example, an employee in the act of helping may move something in a workplace, may bring something into a workspace, may change a setting on equipment, or may inaccurately perform a task. While the intention in each case is to help, in that the helper assumes their action is helping, the action may in fact inadvertently create a hazard for other employees. This may be particularly likely if other employees are not aware of what has been done. However, even when employees know that an employee has helped, a hazard might still result. Consider again the task of chaining down a load of logs on a truck or trailer unit. Assume the employee asks whether they can help. Also assume the other employee accepts the help. A chain is thrown over the load (no one is injured), and the helper attempts to secure their end of the chain and indicates they have done this. The employee on the other side of the truck levers the chain to tighten it. Under this pressure, the chain springs back and the employee tightening the chain is injured. In this example, the helper created a hazard due to a failure to appropriately secure their end of the chain (even though they had indicated they had done this).

8.5.4 The Time Pressure Mechanism

Finally, there is a time element involved in helping another employee which results in time being taken away from the helper's primary job. As such, engaging in helping behaviors may create time pressure for the helper in relation to completing their work. In support of this is Bolino and Turnley's (2005) study which found that organizational citizenship behaviors can increase role overload. Working under time pressure may result in the employee having to rush to complete their work, and this may result in a safety risk. Christian et al.'s (2009) meta-analyze identified pressure from workload as a variable in workplace safety. It is also possible that having to rush to complete ones work provides circumstances where safety rules are not followed.

8.6 Why Help

Clearly, helping can at times be dangerous. Furthermore, Burt et al. (2014) clearly showed that new employees engage in more helping than senior employees. This result may indicate that people do indeed learn from their mistakes, and as such employees' engagement in helping declines as they come to see the associated safety risks. Chapter 3 discussed research on new employee safety expectations. Safety expectations can be unrealistic, particularly in the case of new employees with little or no workplace experience. If work-related safety risks are being inaccurately or unrealistically perceived, or there is an inaccurate perception that systems will ensure safety, a new employee may engage in helping and expose themselves to risks which they simply were not expecting. Alternatively, a new employee may engage in what has been termed *pro-social rule breaking* (Morrison 2006), where the employee clearly knows their behavior (their helping) is a violation of safety rules (is risky), yet they do it anyway.

A number of papers have been published which address rule violations (e.g., Hopkins 2011; Lehman and Ramanujam 2009; Otsuka et al. 2010). Reasons vary for why employees decide not to follow a specific rule or policy. For example, the employee may consider the rule to be wrong, or they may consider that they did not have time to follow the rule. As far as I can determine, the literature on rule breaking has not attempted to isolate motives for rule breaking which could be specifically associated with new employees. However, there are likely to be several mechanisms operating during an employee's initial period of employment which motivate them to help others (and in doing so potentially violate safety rules and policy). For example, new employees may be very enthusiastic to demonstrate their commitment to a job in order to gain job security, and this may result in their engagement in helping behaviors. In the following sections, several factors which may motivate a new employee to engage in helping are examined: need for job security, social integration, need for affiliation, need for respect, and gratitude reciprocity.

8.6.1 Need for Job Security

Arguably, an employee may not be secure in their job until they have proven to the organization that they can perform to a satisfactory level. Furthermore, in some countries, employment law provides for an initial probationary period during which the employee's suitability for permanent employment is being assessed. For example, in New Zealand, the Employment Relations Act (2000) provides for a probation period (normally 3 months) during which an individual's employment can be terminated with relative ease. Thus, both employment law, and an individuals need to have a feeling of job security, could prompt engagement in helping behaviors as a way of demonstrating commitment to the job and the organization. That is, a new employee may be actively looking for every opportunity to demonstrate to the employer that they are a satisfactory employee, and they may see helping others as a positive thing to do.

8.6.2 Social Integration

The arrival of a new employee into a workplace is often the result of employee turnover. A reasonable body of research has examined the impact of turnover on team functioning (See Van der Vegt et al. 2010 for a useful overview). One area that is disturbed when employees leave a team is the social integration of the work group (O'Reilly et al. 1989). There are many positive outcomes associated with a socially integrated team, including gains in performance, trust, and cooperation. In order to reestablish (or establish) social integration when a new employee arrives, members of a work team and the new employee are likely to engage in a range of behaviors, including helping behaviors. Because helping is a pro-social act, it has the ability to help with the formation of social bonds, and thus it has the ability to facilitate social integration.

8.6.3 Need for Affiliation

Individuals with a high need for affiliation have a desire to develop and maintain favorable relationships with others (Greenhalgh and Gilkey 1993). It seems reasonable to predict that individuals needing affiliation with others will engage on helping behaviors to demonstrate their desire for a cooperative and conflict-free relationship. Unfortunately, need for affiliation is a dispositional (personality) variable and as such is difficult for an organization to manage. It is however important to note that need for affiliation is yet another motivator which may drive a new employee to engage in helping acts.

8.6.4 Respect

There is ample research evidence showing that perceived respect, where a team member perceives that other team members respects them, has a positive influence on team performance (e.g., Blader and Tyler 2009; Ellemers et al. 2011; Prestwich and Lalljee 2009). As such, organizations will likely gain a degree of performance advantage by having teams made up of members that respect each other. There is also a very close relationship between liking and respecting (Prestwich and Lalljee 2009). Arguably, people need or desire to be respected and liked, and this can motivate individuals to engage in helping behaviors which lays the foundations for respect development.

Given that respects can focus on a number of different dimensions, including a person's abilities, achievements, determination, and industriousness, it is not surprising that there is a relation between respect and helping. Evidence of this relationship was found by Blader and Tyler (2009) who found a positive correlation between respect and extra-role behavior. In the context of a new employee joining a work team, they are likely to desire the respect of the other team members, and in order to gain this respect, the new employee is likely to engage in helping behaviors which they may see as an opportunity to demonstrate their abilities, industriousness, and enthusiasm.

8.6.5 Gratitude

Help can be provided in a number of forms, including the provision of information and resources, and can result in a feeling of gratitude (Weinstein et al. 2010). It is very likely that a new employee will receive varying amounts of 'help,' from both supervisors and co-workers when they start on the job. For example, Chap. 7 outlines how *compensatory behavior* is necessary for the new employee to adapt, and to gain familiarity in their new job. Depending on how the new employee perceives the motivation for help provided to them, they are likely to feel varying levels of gratitude for the help. For example, if a supervisor shows a new employee how to fill in a particular form, this may be perceived as part of the supervisor's job, and not result in feelings of gratitude. In contrast, if the new employee's co-workers show the new employee how to perform aspects of the job, and this is *not formally* required of the co-workers, this may occasion feelings of gratitude. Thus, a key aspect for a helping interaction to result in the generation of gratitude is the perception that the helping act was a favor and was not required to be given (Bartlett and DeSteno 2006).

Gratitude produced from helping is likely to result in a number of positive outcomes, including positive relationships and trust between the helper and recipient (Algoe et al. 2008). Gratitude can also be a motivator in that there is a reciprocal aspect to gratitude (Grant and Gino 2010; Rotkirch et al. 2014). Basically, 'you helped me, so I will/should help you.' Gratitude also appears to extend beyond the initial dyadic relationship, in that it seems to motivate the helping of others, 'I was helped, therefore I will/should help others' (Chang et al. 2012). Furthermore, gratitude may prompt helping which is actually costly to the helper (Bartlett and DeSteno 2006). Clearly, gratitude has a strong motivating component, and new employees experiencing gratitude are likely to engage in reciprocal helping behaviors.

While reciprocating helping will strengthen positive relationships and trust in a work group, there are too many risks associated with new employee helping and they should be encouraged not to reciprocate when help is given. It will be important to instruct new employees not to reciprocate when help is given to them, and explain why in terms of the issues discussed above. Of course, a better strategy is to limit the generation of gratitude by ensuring a formal relationship between a new employee and any one that provides them with help (e.g., co-workers), such that any help they provide the new employee is perceived by the new employee as just part of the helpers job.

8.7 Managing New Employee Helping Behaviors

Clearly, there may be safety risks associated with helping behaviors. Unfortunately, some of the factors that can motivate an employee to engage in helping (e.g., job security, development of social integration, respect, and gratitude generated reciprocal helping) are particularly associated with new employees. As such, helping-related accidents are a particular concern for new employees. As noted in Chap. 7 of this book, new employees will take some time to adapt to, and familiarize with, aspects of their new job. Adding safety risks associated with helping into the initial employment period is unnecessary. Thus, an organization should develop a strategy to manage new employee helping behaviors.

The topic of helping in the workplace can be discussed during new employee induction. However, simply instructing a new employee to stay within the boundaries of their job may not be sufficient to stop them engaging in helping. While this instruction (given during induction and reinforced by supervisors) should be given, it must be complimented with other strategies. During induction, new employees can be asked to complete the consequences of helping scale (see Chap. 9, Sect. 9.3). Completion of this scale, and a consideration of how the new employee responds, should help clarify for the new employee the safety implications for both themselves and the employee being helped. In addition to this focus on helping risks during new employee induction, an organization can adopt two other strategies to help manage new employee helping. The first is to develop *a safety conscious helping culture* within the organization. The second is to train new employees in *a think before you help process*, which allows them to evaluate each situation where they may consider helping. Each of these strategies is discussed below.

8.7.1 A Safety Conscious Helping Culture

Clearly, organizations can benefit from a workforce that engages in organizational citizenship behaviors, part of which is helping. One might even characterize a workforce where organizational citizenship behaviors are the norm as having a *performance conscious helping culture*. However, it is clear that there are times, and situations, when helping is to be discouraged, if performance is to be maintained. That is an accident not only has consequences in terms of injury and personal trauma, but also costs the organization and effects performance outcomes. I would argue that a *safety conscious helping culture* should be based on 8 principles: (1) recognizes when helping is not appropriate—by adopting a *think before you help process*, (2) recognizes the value of not blindly engaging in helping behavior, (3) promotes respect for employees that are careful about how and when they help, (4) encourages gratitude for behavioral restraint, (5) reassures new employees that not helping will have only positive consequences, (6) indicates that not helping will not be seen as a lack of commitment to the job or organization, (7) recognizes that not helping is not a sign of a lack of gratitude for the help others may have given, and (8) not helping will increase, rather than decrease, respect from fellow workers, and respect will be given for the care taken by the new employee to ensure everyone's safety.

8.7.2 A Think Before You Help Process

Helping in a real-time sense is an individual making a decision, followed by a course of action. While an organization may have clearly communicated a desire for a safety conscious helping culture, this is not necessarily going to be adopted by all employees and/or teams within an organization. The vast literature on safety culture versus safety climate clearly shows how individuals' behavior can vary from an organization's safety culture principles. Thus, employees need to be trained to evaluate situations in which they are considering helping with the objective of ensuring that helping is only undertaken if it safe to do so. Table 8.2 shows a

Table 8.2 Key questions to consider before engaging in helping

Helping safety issue	Key question	Don't help
Why	Why are you going to help?	Don't know
Task safety risk	Does the task have safety risks?	Yes
Environment safety risk	Does the environment where the task is being performed add safety risks?	Yes
Helper ability	Do I have the knowledge, skills, and abilities to help safely?	No
Communication: acknowledge	Have I asked if I can help?	No
Absence effects	Could there be a safety issue if I leave my job to help?	Yes
Hazard creation potential	Could my helping potentially create a hazard?	Yes
Forgetting effects potential	Could the interruption created by helping make me forget something important?	Yes
Time pressure	Could taking time to help increase my workload, or make me cut-corners, on my job?	Yes

number of key questions which the employee should answer before helping. If any question produces the answer shown in the right-hand column of Table 8.2, helping should not be undertaken.

8.8 Conclusions

Acts of helping can have many positive outcomes. At the organizational level, they can increase performance, and at the individual level they can increase employee social integration, and develop trust and respect between co-workers. The aim of this chapter is not to argue that helping should not be encouraged. However, it is clear that there are many safety risks associated with helping. Furthermore, it is clear that new employees are perhaps the least well equipped to deal with the safety risks associated with helping, yet many of the factors which motivate helping are likely to be particularly associated with new employees. New employees should not blindly engage in helping, and organizations should put in place a *safety conscious helping culture* and train all employees in a *think before you help process*.

The chapter has hopefully not only highlighted the risks associated with helping but also indicated two strategies which might help reduce helping associated accidents. It is important to note that the motivators of helping are likely to be very strong, and simply asking new employees to refrain from helping is not likely to be entirely successful. Developing a safety conscious helping culture should help as it will put the key issues into a forum where they can be discussed, and hopefully the key principles I have suggested for a safety conscious helping culture will become the workplace norm. It is also important to train employees to engage in a think before you help process. Appropriate application of the questions shown in Table 8.2, and a refrain from helping when the circumstances dictate, should vastly reduce helping associated accidents for all employees.

References

Algoe, S. B., Haidt, J., & Gable, S. L. (2008). Beyond reciprocity: Gratitude and relationships in everyday life. *Emotion, 8*(3), 425–429.

Bartlett, M. Y., & DeSteno, D. (2006). Gratitude and prosocial behavior: Helping when it costs you. *Psychological Science, 17*(4), 319–325.

Bell, S. J., & Menguc, B. (2002). The employee-organization relationship, organizational citizenship behaviors and superior service quality. *Journal of Retailing, 78*, 131–146.

Blader, S. L., & Tyler, T. R. (2009). Testing and extending the group engagement model: Linkages between social identity, procedural justice, economic outcomes, and extrarole behavior. *Journal of Applied Psychology, 94*(2), 445–464.

Bolino, M. C., & Turnley, W. H. (2005). The personal costs of citizenship behavior: The relationship between individual initiative and role overload, job stress, and work-family conflict. *Journal of Applied Psychology, 90*, 740–748.

Borman, W. C., & Motowildo, S. J. (1997). Task performance and contextual performance: The meaning for personnel selection research. *Human Performance, 10*(2), 99–109.

Burt, C. D. B., Banks, M., & Williams, S. (2014). The safety risks associated with helping others. *Safety Science, 62*, 136–144.

Carpenter, N. C., Berry, C. M., & Houston, L. (2014). A meta-analytic comparison of self-reported and other-reported organizational citizenship behavior. *Journal of Organizational Behavior, 35*, 547–574.

Chang, Y., Lin, Y., & Chen, L. (2012). Pay it forward: Gratitude in social networks. *Journal of Happiness Studies, 13*, 761–781.

Chiaburu, D. S., Oh, I., Berry, C. M., Li, N., & Gardner, R. G. (2011). The five-factor model of personality traits and organizational citizenship behaviors: A meta-analysis. *Journal of Applied Psychology, 96*(6), 1140–1166.

Christian, M. S., Bradley, J. C., Wallace, J. C., & Burke, M. J. (2009). Workplace safety: A meta-analysis of the role of person and situation factors. *Journal of Applied Psychology, 94*, 1103–1127.

Cohen, H. H., Templer, J., & Archea, J. (1985). An analysis of occupational stair accident patterns. *Journal of Safety Research, 16*, 171–181.

De Jong, S. B., Van der Vegt, G. S., & Molleman, E. (2007). The relationship among asymmetry in task dependence, perceived helping behaviour, and trust. *Journal of Applied Psychology, 92*(6), 1625–1637.

Didla, S., Mearns, K., & Flin, R. (2009). Safety citizenship behaviour: A proactive approach to risk management. *Journal of Risk Research, 12*(3–4), 475–483.

Dyne, L., & LePine, J. A. (1998). Helping and voice extra-role behaviors: Evidence of construct and predictive validity. *Academy of Management Journal, 41*(1), 108–119.

Ellemers, N., Sleebos, E., Stam, D., & de Gilder, D. (2011). Feeling included and valued: How perceived respect affects positive team identity and willingness to invest in the team. *British Journal of Management, 24*, 21–37.

Grant, A. M., & Gino, F. (2010). A little thanks goes a long way: Explaining why gratitude expressions motivate prosocial behavior. *Journal of Personality and Social Psychology, 98*(6), 946–955.

Greenhalgh, L., & Gilkey, R. (1993). The effect of relationship orientation on negotiators' cognitions and tactics. *Group Decision and Negotiation, 2*, 167–183.

Gyekye, S. A., & Salminen, S. (2005). Are "good soldiers" safety conscious? An examination of the relationship between organizational citizenship behaviors and perceptions of workplace safety. *Social Behavior and Personality, 33*, 805–820.

Hoffman, B. J., Blair, C. A., Meriac, J. P., & Woehr, D. J. (2007). Expanding the criterion domain? A quantitative review of the OCB literature. *Journal of Applied Psychology, 92*, 555–566.

Hofmann, D. A., Morgeson, F. P., & Gerras, S. J. (2003). Climate as a moderator of the relationship between leader-member exchange and content specific citizenship: Safety climate as an exemplar. *Journal of Applied Psychology, 88*, 170–178.

Hopkins, A. (2011). Risk-management and rule-compliance: Decision-making in hazardous industries. *Safety Science, 49*, 110–120.

Jiag, X., Master, R., Kelkar, K., & Gramopadhye, A. K. (2002). Task analysis of shift activity in aviation maintenance environment: Methods and findings. *Human Factors and Aerospace Safety, 2*(1), 45–69.

Jiao, C., Richards, D. A., & Hackett, R. D. (2013). Organizational citizenship behavior and role breath: A meta-analytic and cross-cultural analysis. *Human Resource Management, 52*(5), 697–714.

Lehman, D. W., & Ramanujam, R. (2009). Selectivity in organizational rule violations. *Academy of Management Review, 34*(4), 643–657.

Matric, J., Davidson, P. M., & Salsmonson, Y. (2010). Review: Bringing patient safety to the forefront through structured computerization during clinical handover. *Journal of Clinical Nursing, 20*, 184–189.

Mayor, E., Bangerter, A., & Aribot, M. (2011). Task uncertainty and communication during nursing shift handovers. *Journal of Advanced Nursing, 68*(9), 1956–1966.

Morrison, E. W. (2006). Doing the job well: An investigation of pro-social rule breaking. *Journal of Management, 32*, 5–28.

Naumann, S. E., & Ehrhart, M. G. (2011). Moderators of the relationship between group helping norms and individual helping. *Small Group Research, 42*, 225–248.

Nielsen, T. M., Hrivnak, G. A., & Shaw, M. (2009). Organizational citizenship behavior and performance: A meta-analysis of group-level research. *Small Group Research, 40*, 555–577.

Norman, D. A. (1981). Categorization of action slips. *Psychological Review, 88*, 1–15.

O'Reilly, C. A, I. I. I., Caldwell, D. F., & Barnett, W. P. (1989). Work group demography, social integration, and turnover. *Administrative Science Quarterly, 34*, 21–37.

Oosterhof, A. Van, der Vegt, G., Van de Vliert, E., & Sanders, K. (2009). Valuing skill differences: Perceived skill complementarity and dyadic helping behavior in teams. *Group and Organization Management, 34*, 536–562.

Organ, D. W. (1988). *Organizational citizenship behavior: The good soldier syndrome.* Lexington, MA: Lexington.

Organ, D. W., Podsakoff, P. M., & MacKenzie, S. B. (2006). *Organizational citizenship behavior: Its nature.* Thousand Oaks, CA, Sage: Antecedents and Consequences.

Organ, D. W., & Ryan, K. (1995). A meta-analytic review of attitudinal and dispositional predictors of organizational citizenship behavior. *Personnel Psychology, 48*, 775–802.

Otsuka, Y., Misawa, R., Noguchi, H., & Yamaguchi, H. (2010). A consideration for using workers heuristics to improve safety rule based on relationships between creative mental sets and rule-violating actions. *Safety Science, 48*, 878–884.

Podsakoff, P. M., & Mackenzie, S. B. (1994). Organizational citizenship behavior and sales unit effectiveness. *Journal of Marketing Research, 31*, 351–363.

Podsakoff, P. M., Mackenzie, S. B., Paine, J. B., & Bachrach, D. G. (2000). Organizational citizenship behaviors: A critical review of the theoretical and empirical literature and suggestions for future research. *Journal of Management, 26*, 513–563.

Prestwich, A., & Lalljee, M. (2009). The determinants and consequences of intragroup respect: An examination within a sporting context. *Journal of Applied Social Psychology, 39*(5), 1229–1253.

Rotkirch, A., Lyons, M., David-Barrett, T., & Jokela, M. (2014). Gratitude for help among adult friends and siblings. *Evolutionary Psychology, 12*(4), 673–686.

Spector, P. E., & Che, X. X. (2014). Re-examining citizenship: How the control of measurement artifacts affects observed relationships of organizational citizenship behavior and organizational variables. *Human Performance, 27*, 165–182.

Tjosvold, D., Yu, Z., & Hui, C. (2004). Team learning from mistakes: The contribution of cooperative goals and problem-solving. *Journal of Management Studies, 41*(7), 1223–1245.

Tremblay, S., Vachon, F., Lafond, D., & Kramer, C. (2012). Dealing with task interruptions in complex dynamic environments: Are two heads better than one? *Human Factors, 54*(1), 70–83.

Van der Vegt, G. S., Bunderson, S., & Kuipers, B. (2010). Why turnover matters in self-managing work teams: Learning, social integration, and task flexibility. *Journal of Management, 36*(5), 1168–1191.

Van der Vegt, G. S., Bunderson, J. S., & Oosterhof, A. (2006). Expertness diversity and interpersonal helping in teams: Why those who need the most help end up getting the least. *Academy of Management Journal, 49*(5), 877–893.

Venkataramani, V., & Dalal, R. S. (2007). Who helps and harms whom? Relational antecedents of interpersonal helping and harming in organizations. *Journal of Applied Psychology, 92*(4), 952–966.

Wallace, J. C., & Chen, G. (2005). Development and validation of a work-specific measure of cognitive failure: Implications for occupational safety. *Journal of Occupational and Organizational Psychology, 78*, 615–632.

Wallace, J. C., & Vodanovich, S. J. (2003). Can accidents and industrial mishaps be predicted? Further investigation into the relationship between cognitive failure and reports of accidents. *Journal of Business and Psychology, 17,* 503–514.

Weinstein, N., DeHann, C. R., & Ryan, R. M. (2010). Attributing autonomous versus introjected motivation to helpers and the recipient experience: Effects on gratitude, attitudes, and well-being. *Motivation and Emotion, 34,* 418–431.

Whitman, D. S., Van Rooy, D. L., & Viswesvaran, C. (2010). Satisfaction, citizenship behaviors, and performance in work units: A meta-analysis of collective construct relations. *Personnel Psychology, 63,* 41–81.

Chapter 9
Measuring New Employee Safety-Related Variables

9.1 Introduction

There are many scales that have been developed to measure safety-related variables. The majority of these focus on aspects of safety climate. It is not the intention of this chapter to examine these measures. Rather, the specific focus is on the factors which are directly related to new employee safety. Thus, the measures discussed in this chapter are restricted to those which measure attitudes and expectations which new employees bring to the workplace; worker attitudes and behaviors which are particularly important for new employee adaption; and behaviors, such as helping, which are associated with being a new employee. It is the opinion of this author that measurement provides evidence which can be presented to new employees, co-workers, and management in order to help explain the safety issues associated with new employees. Furthermore, the collection of data provides a degree of precision in terms of the issues faced by a specific organization, for a specific job, and related to the type of new employees being recruited.

For scales to measure other safety-related factors, the reader can consult Costa and Anderson (2011) for trust measures; Zohar (2000) for safety climate measures; Barling et al. (2002) for safety consciousness; Sneddon et al. (2013) for situational awareness; Neal and Griffin (2006) for safety participation and compliance; Chmiel (2005) for bending the rules; Cox and Cox (1991) for safety skepticism; Neal et al. (2000) for safety knowledge and safety motivation; Tucker et al. (2008) for employee safety voicing; Tucker et al. (2008) for perceived organizational and perceived co-worker support for safety; and Diaz-cabera et al. (2007) for safety culture. Another good source of information on safety measures are meta-analyses (e.g., Christian et al. 2009; Clarke 2006).

The scales and measurement options discussed in this chapter are presented in the order which they might be used to manage new employee safety. Section 9.2 examines measures of new employee safety expectations, Sect. 9.3 examines measures which provide an awareness of helping safety risks, and Sect. 9.4

© Springer International Publishing Switzerland 2015
C.D.B. Burt, *New Employee Safety*,
DOI 10.1007/978-3-319-18684-9_9

discusses a safety exit survey process which can be used to gain information about a job's safety risk profile. As discussed in Chap. 3, new employees can have safety expectations which vary considerably. When safety expectations are unrealistic, they can expose a new employee to risk and the possibility of an accident increases. Before a new employee starts their job, it is vital that they have realistic safety expectations. In a similar way, a job's safety risks can range from the normal and expected level, to extreme and unexpected. If a job's safety risk profile has for some reason extended beyond what is normally associated with the type of work, either corrective action needs to be taken, or the new employee needs to be warned of the these circumstances. Thus, information on both new employee safety expectations and job safety risks needs to be collected, and where necessary interventions undertaken if new employee safety is to be ensured.

9.2 New Employee Safety Expectations

Chapter 3 discussed research on safety expectations. New employees will hold expectations for a range of different aspects related to safety in the job they are entering. These expectations will relate mainly to the behavior of co-workers and management, but new employees will also have expectations about their own behavior. Research by Burt et al. (2012) demonstrated that safety expectations can be unrealistic. Of course a new employee does not know that their expectations about safety may be unrealistic, nor that a mismatch between what they expect to occur and what really is going to occur in their new job could expose them to safety risks. In order to provide new employees with this understanding, what they expect across a range of variables needs to be compared with what exiting employees indicate is the organizational safety reality. Sections 9.2.1 to 9.2.5 show measures of expectations related to co-workers: *expected familiarization by co-workers*, *expected co-worker safety communication, expected co-worker safety behavior, and expected reactions to new employees*. Section 9.2.6 covers expectations related to management, Sect. 9.2.7 supervision, and Sect. 9.2.8 a scale to measure a new employee's expectation of their behavior: *new employee expected behavior*.

In order for new employees' safety expectations to be assessed, they need to be compared with current job incumbents' perceptions of the relevant constructs. Thus, the scale shown in each section has a version which the new employee is asked to complete and a version which job incumbents are asked to complete. Slightly different instructions are provided for new employees (i.e., *These questions are about your expectations of ... in the job you are about to start. For each statement, please circle the number which indicates the extent to which you disagree or agree.*) and job incumbents (i.e., *These questions are about ... in your workplace. For each statement, please circle the number which indicates the extent to which you disagree or agree*). The scale title is added into the blank space in the instructions. For each scale, items are rated on a 7 point Likert scale where 1 = strongly disagree, 4 = neither disagree or agree, and 7 = strongly agree.

Expectation scales (both new employee and incumbent versions) can be scored by summing the rating for each item and dividing the sum by the number of scale items. The overall scale score (the average rating) can be presented to a new employee (for an example of a feedback table format, see Table 3.5 in Chap. 3). However, more safety advantage may come from feeding back ratings for the individual scale items and in particular by providing new employees with a comparison between their ratings and the average rating for each scale item given by job incumbents. The key objective is to look for items where the new employee and incumbents ratings are significantly different (particularly where the new employee is giving a larger rating than the average obtained from job incumbents). In the case where a new employee is clearly holding expectations which are not consistent with job incumbents, there should be a discussion with the new employee about the implications of their expectations for their, and their co-workers, safety. For example, if a new employee's responses indicate that they have overestimated how much co-workers will ensure their safety (have an unrealistic view of this)—what does this mean for their safety.

9.2.1 Expected Co-worker Behavior

Expected co-worker behavior is measured by 4 scales relating specifically to familiarization, and safety communication, and more generally to safety behaviors and reactions to new employees. Use of all 4 scales will help clarify a new employee's expectations about co-workers and also help inform new employees about the way co-workers are behaving in their new workplace. Tables 9.1, 9.2, 9.3 and 9.4 show the items from each scale. It is important to note that Chap. 8 which dealt with helping behaviors outlined the safety risks which can be associated with helping generated through gratitude associated reciprocity. It was argued in Chap. 8 that co-workers should be careful about providing random acts of helping to new employees, as this can begin a cycle where a new employee feels they need to pay back the help. However, new employees are likely to have a general expectation that co-worker will help them adjust during their initial period of employment. As such, it is particularly important to measure new employee co-worker related safety expectations and ensure that new employees have realistic expectations around co-worker behavior.

9.2.2 Expected Familiarization by Co-workers

New employees are likely to hold expectations about the degree to which co-workers will help them become familiarized with aspects of the workplace. As discussed in Chap. 7, new employees go through a period of adaption when they enter a new job, and part of the process of adapting is the acquisition of familiarity

Table 9.1 The expected familiarization by co-workers scale items

New employee items	Job incumbents items
Members of my workplace will familiarize me with the specific characteristics of the equipment which they use	Members of my workplace familiarize new employees with the specific characteristics of the equipment which they use
Members of my workplace will familiarize me with the specific characteristics of the physical environments within which they work	Members of my workplace familiarize new employees with the specific characteristics of the physical environments within which they work
Members of my workplace will familiarize me with the specific operational procedures which they use	Members of my workplace familiarize new employees with the specific operational procedures which they use
Members of my workplace will familiarize me with the specific way in which they do their job	Members of my workplace familiarize new employees with the specific way in which they do their job

Table 9.2 The expected safety communication scale items

New employee items	Job incumbents items
Co-workers will discuss changes that could improve safety	Co-workers discuss changes that could improve safety
Co-workers will give each other informal safety instruction	Co-workers give each other informal safety instruction
Co-workers will discuss near hits	Co-workers discuss near hits
Co-workers will discuss past accidents	Co-workers discuss past accidents
Co-workers will remind each other of the need to follow safety regulations	Co-workers remind each other of the need to follow safety regulations
Co-workers will say a good word whenever they see a job done according to the safety rules	Co-workers say a good word whenever they see a job done according to the safety rules
Co-workers will approach each other during work to discuss safety issues	Co-workers approach each other during work to discuss safety issues
Co-workers will point out hazards to co-workers	Co-workers point out hazards to co-workers
Co-workers will notify crew leaders of hazards	Co-workers notify crew leaders of hazards
Co-workers will report accidents and near misses to management	Co-workers report accidents and near misses to management

with the job's equipment, the work environment, the operational procedures, and with co-worker behavior. A new employee who is expecting to get more help with these aspects of familiarization from co-workers than co-workers in this job typically give may be at risk, as they are essentially assuming that co-workers are going to provide information and clarity, when in fact this is perhaps not going to happen. If this is the case, a new employee may assume they have been told what they need to know, when in fact key information has not been provided. As noted, Chap. 8 outlined several reasons why co-workers should not give help to new employees,

Table 9.3 The expected co-workers reactions to new employees' scale items

New employees	Job incumbents
Co-workers will pay more attention to safety when a new employee joins	Co-workers pay more attention to safety when a new employee joins
Co-workers will encourage a new employee to ask about safety procedures	Co-workers encourage a new employee to ask about safety procedures
Co-workers will immediately determine the safety attitudes of a new employee	Co-workers immediately determine the safety attitudes of a new employee
Co-workers will find out the safety history of a new employee	Co-workers find out the safety history of a new employee

Table 9.4 The expected co-worker safety behavior scale items

New employee	Job incumbents
Co-workers will warn each other when their actions are unsafe	Co-workers warn each other when their actions are unsafe
Co-workers will assist each other with tasks to ensure safety	Co-workers assist each other with tasks to ensure safety
Co-workers will recognize each others' limitations	Co-workers recognize each others' limitations
Co-workers will expect other workers to behave safely	Co-workers expect other workers to behave safely
Co-workers who work safely will try to emphasize it and make sure others do the same	Co-workers who work safely emphasize it and make sure others do the same
Co-workers will immediately remove hazards if possible	Co-workers immediately remove hazards if possible

unless the relationship is <u>formally</u> established as part of the co-workers job. It is very important that new employees understand this and also why such a policy is in place.

The *expected familiarization by co-workers scale* has 4 items, and both versions are shown in Table 9.1. At this time, the scale has not been used in published research. Unpublished analysis of 144 new employee responses produced a Cronbach's alpha of 0.77.

9.2.3 Expected Co-worker Safety Communication

Information about safety is vital to ensure employee safety. All employees should expect to receive information about safety, and research has clearly shown that communication about safety reduces accidents. The expected safety communication scale has items which examine different aspect of safety communication. A new employee may be expecting such communication, and it is important that new employees understand the degree to which job incumbents engage in safety communication. As with all interactions between co-workers and new employees, ensuring that safety communication is a formal part of all employees job will reduce the problems associated with helping generated through gratitude associated reciprocity.

The *expected safety communication scale* has 10 items, and both versions are shown in Table 9.2. At this time, the scale has not been used in published research. Unpublished analysis of 144 new employee responses produced a Cronbach's alpha of 0.79.

9.2.4 Expected Co-workers Reactions to New Employees

Co-workers should realize that new employees are a safety risk, and pose a danger both to themselves and to other workers. A safety advantage may be gained by workers responding in a positive (safety conscious way) when a new employee joins a workplace. Furthermore, new employees may be expecting that co-workers will actively engage with them to ensure their (and others) safety.

The expected co-workers reactions to new employees' scale has 4 items, and both versions are shown in Table 9.3. The items for the new employee version were adapted (by rewording into future tense) from scales developed by Burt and Stevenson's (2009) and Burt et al.'s (2009). Burt et al.'s (2012) reported an alpha value of 0.86 for the new employee version, and 0.76 for the job incumbent version.

9.2.5 Expected Co-worker Safety Behavior

The expected co-worker safety behavior scale has 6 items, and both versions are shown in Table 9.4. Scale items are adapted (by rewording into future tense) from Burt et al. (1998) *CARE* scale, and from Mueller et al. (1999) *Co-worker commitment to safety* scale. Burt et al. (2012) reported Cronbach's alphas of 0.84 and 0.83 for the new employee version, and 0.81 for the job incumbent version.

9.2.6 Expected Management Safety Behavior

Organizations vary considerably in terms of how they manage safety. Thus, the expectations of management safety behavior formed from one workplace may have little basis in reality in another workplace. At this point, it is also worth noting the vast literature on safety culture and safety climate. Safety culture stems from the organization and is the top-down safety values, beliefs, and norms, while safety climate is more accurately defined as the employee's perceptions of how various aspects of the working environment impact on their safety (see Bjerkan 2010, for a more detailed discussion of the relationship between safety culture and climate, and its impact on team safety). From the point of view of this section, it is sufficient to understand that an organization's safety culture (and all that it entails) may be viewed differently by different teams. When a team collectively perceives safety in the same way as the organization (assuming a positive perception), the team might be said to have a strong or positive safety climate. Furthermore, this situation should make the team a safer option for the integration of a new employee. Completion of the *expected management safety behavior scale* provides for a new employee to understand how co-workers view management's approach to safety.

A new employee will rely to some extent on management to ensure their safety, and may monitor and safeguard less when they assume that management is actively engaged with safety management. Thus, like other areas, it is important that the reality of management's safety behavior (at least as perceived by current job incumbents) is what the new employee is expecting. By comparing their expected management safety behavior scale ratings to those of job incumbents, the new employee should quickly realize where their expectations are out of sink with the realities of the particular workplace.

The expected management safety behavior scale used by Burt et al. (2012) has 13 items, and both versions are shown in Table 9.5. Scale items were adapted from Chmiel's (2005) *management safety climate scale*, and from Walker and Hutton's (2006), scale measuring how management deal with safety. Burt et al. (2012) reported Cronbach's alphas for the new employee version of 0.92 and 0.88, and a value of 0.89 for the incumbent version.

9.2.7 Expected Supervision

The expected supervision scale has 6 items, and both versions are shown in Table 9.6. Scale items were developed based on the discussion of supervisor behavior required to ensure new employee safety in Chap. 4, Sect. 4.2.8. At the time of writing, no data on the psychometric properties of this scale had been collected. As noted in Chap. 3, supervision of new employees should be a specific task assigned to supervisors. Furthermore, new employees are likely to expect that supervisors will be there to ensure their safety. As noted in many places in this book, the perception that a system has a component which is there to protect a person from risk can lead to more risk being taken. Thus, it is very important that new employees have a realistic perception of the degree of supervision that they will receive. It is also important to note that employees (job incumbents) are asked to complete this scale—not supervisors. Employees should be able to respond to the items in terms of the experiences they have had with supervision, whereas supervisors may respond in terms of what higher management expect of them, rather than their actual supervision of new employees.

Table 9.5 The expected management safety behavior scale items

New employee	Job incumbent
Management will be quick to respond to the safety concerns of employees	Management are quick to respond to the safety concerns of employees
Management will be actively involved in safety programs	Management are actively involved in safety programs
Management will maintain a safe workplace	Management maintain a safe workplace
Management will take a proactive approach to safety	Management take a proactive approach to safety
Management will conduct regular safety training with all employees	Management conduct regular safety training with all employees
Management will make sure that work demands do not compromise safety	Management make sure that work demands do not compromise safety
Management will regularly update safety documentation	Management regularly update safety documentation
Management will supply enough resources to get the job done safely	Management supply enough resources to get the job done safely
Management will ensure that employees can attend safety training sessions	Management ensure that employees can attend safety training sessions
Management will inform employees about new safety rules	Management inform employees about new safety rules
Management will communicate the organization's safety objectives to all employees	Management communicate the organization's safety objectives to all employees
Management will set a good example for safety behavior	Management set a good example for safety behavior
Management will carry out regular safety inspections	Management carry out regular safety inspections

Table 9.6 The expected supervision scale items

New employees' items	Job incumbents items
Supervisors will pay more attention to safety when a new employee joins	Supervisors pay more attention to safety when a new employee joins
Supervisors will encourage a new employee to ask about safety procedures	Supervisors encourage a new employee to ask about safety procedures
Supervisors will give more attention to new employees than to other employees	Supervisors give more attention to new employees than to other employees
Supervisors will understand that new employees need help familiarizing with the job, equipment, and work procedures	Supervisors understand that new employees need help familiarizing with the job, equipment, and work procedures
Supervisors will ensure new employees can safely perform tasks assigned to them	Supervisors ensure new employees can safely perform tasks assigned to them
Supervisors will relax performance expectations for new employees to ensure their safety	Supervisors relax performance expectations for new employees to ensure their safety

9.2.8 New Employee Expected Behavior

It is also important that a new employee understands how they will behave when they enter a job. Job incumbents will, to varying degrees, have experienced new employees in the past, and these individuals will have varied in terms of their safety behavior. New employee behavior will be partly determined by what they know, partly determined by how they see others behaving, and partly determined by their motivation to be engage with safety. Providing new employees with feedback on how they expect to behave in relation to safety, and comparing this with how job incumbents are assuming they will behave should produce a safety advantage. That is, a new employee may say they will engage with safety, yet it might be the collective opinion of co-workers that this does not always happen. Showing new employees via feedback how co-workers think new employees typically behave, and providing new employees with a comparison with what they have said (how they have rated items) should produce a safety advantage.

The new employee expected behavior scale has 10 items, and both versions are shown in Table 9.7. The items for the new employee version were adapted (by rewording into first person singular future tense) from Chmiel's (2005) work on job safety behavior and Walker and Hutton's (2006) research. Burt et al. (2012) reported Cronbach's alphas for the new employee version of 0.85, and a value of 0.90 for the job incumbent version.

Table 9.7 The new employee expected behavior scale items

New employee items	Job incumbent items
I will be familiar with safety documentation	New employees are familiar with safety documentation
I will maintain a clean, safe, work environment	New employees maintain a clean, safe, work environment
I will inform incoming shifts or work teams of current hazards and risks	New employees inform incoming shifts or work teams of current hazards and risks
I will follow safety rules	New employees follow safety rules
I will take responsibility for safety	New employees take responsibility for safety
I will set an example of safe working behavior	New employees set an example of safe working behavior
I will raise safety concerns	New employees raise safety concerns
I will take a proactive approach to safety	New employees take a proactive approach to safety
I will report safety incidents or near misses in an objective, factual manner	New employees report safety incidents or near misses in an objective, factual manner
I will voluntarily carry out tasks or activities that help to improve workplace safety	New employees voluntarily carry out tasks or activities that help to improve workplace safety

9.3 Awareness of Helping Safety Risks

Burt, Banks, and Williams (2014) reported one of the few studies that has examined the safety risks associated with helping. Chapter 8 provides an extensive discussion of the safety risks associated with helping. Chapter 8 also makes a number of predictions about why new employees may be particularly likely to engage in helping behaviors. Part of the strategy to manage helping behaviors is to make new employees aware of the risks that can be associated with helping. In order to make new employees aware of the potential consequences of helping behaviors, new employees can be asked to complete the items in the *consequences of helping scale* shown in Table 9.8. The *consequences of helping scale* could be completed by new employees during induction.

The items in the *consequences of helping scale* are adapted from the measures used in Burt, et al. (2014). At the time of writing this scale had not been used, and as such no psychometric data are available. New employees would rate each item on a 7 point scale where 1 = strongly disagree and 7 = strongly agree. While self-report responses can be open to social desirability responding, with the *consequences of helping scale,* it is not easy for a person to know what is the correct (socially desirable) answer, should they rate items so as to indicating agreement which implies helping is bad (risky), or should they disagree and imply helping is good. At the very least, the completion of the scale provides for the topic to be discussed with new employees during their induction and provides some data upon

Table 9.8 Items in the consequences of helping scale	Doing something to help another employee which they were not expecting can be risky
	Doing something to help another employee which you have not immediately told them about can be risky
	It is possible to forget at what point you are in your work when returning from helping another employee
	It is possible to forget something you were planning to do after returning from helping another employee
	While helping another employee something unexpected can happen in relation to **your** job
	While helping another employee something unexpected can happened in relation to **their** job
	Doing what you think will be helpful for another employee can turn out to be a safety risk for **you**
	Doing what you think will be helpful for another employee can turn out to be a safety risk for **them**
	Doing what you think will be helpful for another employee can turn out to be a safety risk for **another member of the organization**
	Having to rush to complete your tasks because of spending time helping another employee can be a safety risk

which this discussion can be based. Clearly, from a safety perspective, new employees should agree with the items shown in Table 9.8.

9.4 Job Safety Risk Profile: The Safety Specific Exit Survey

There are number of variables that can be used to measure a job's safety risk profile. For example, accident statistics can be examined for evidence that safety is clearly a factor associated with the job; job analysis work can highlight risky tasks and equipment; and employees can be asked to indicate their perceptions of a job's safety risk. The later approach could use Hayes et al. (1998) 10-item Work Safety Scale to measure perceived job risk. Example items from this scale are *hazardous, dangerous, risky, and chance of death* to which the employee responses on a 5-point Likert scales, anchored with 1 = strongly disagree to 5 = strongly agree. A higher score indicates greater perceived risk in the job.

Each approach to profiling a job's safety risk will provide a unique perspective. Furthermore, collectively the data from these assessments will reflect a job's safety risk profile. Chapter 3 discusses how information on a job's safety risks should be included in a job description document (see Fig. 3.1). For any specific job, the safety risk profile assessment can show a range from a normal and expected safety risk level, through to an extreme and unexpected safety risk. That is, while an organization may think that its management systems are controlling a job's safety risks to a normal and expected level, reality maybe very different. Of course, it would be particularly risky to bring a new employee into a job that has safety risks beyond the normal and expected level.

A further approach to assessing a job's safety risks is to collect safety information from employees that leave the job. As noted in Chaps. 2 and 4, safety issues can result in employees resigning from their work, and as such, job vacancies can be the result of safety issues. Furthermore, there can be issues around employees' voicing safety issues, and these are largely removed when an employee has resigned from a job. Burt et al. (2013) provided a discussion of the use of a safety specific exit survey process as a way of collecting information from resigning employees about a job's safety risks. Items which can be included in a safety specific exit survey are shown in Table 9.9. These items should be applicable to many different types of work; however, they should be examined for relevance before being adopted. Furthermore, each job may have idiosyncratic aspects which can be formed into a safety specific exit survey item. Burt et al. (2013) used instructions similar to these in their use of the safety specific exit survey process: *Listed below are 'safety issues' which you might have wanted to talk to either your co-workers or management about. For each safety issue, please respond by ticking* **one or more boxes:** *tick* **Yes management** *if it was an issue you would have liked to talk to management about but never did;* **Yes co-worker** *if it was an issue you would have liked to talk to co-workers about but never did.* The responses from a resigning employee will help highlight safety issues associated with a job, and may

Table 9.9 Safety specific exit survey items as used in Burt et al. (2013)

Safety issues	
Awareness that new recruits can pose a safety risk	Work speed pressure from supervisors which reduced safety
New recruits being alerted to the risks involved in their job	Too much work to perform safely
New recruits understanding of safety policy	Work-related fatigue which reduced safety
New recruits lack of sufficient experience to work safely	Insufficient staff to complete the job safely
New recruits lack of skills and abilities to work safely	Working methods which decreased safety
New recruits behaving unsafely	Safety policy/rules which seemed to reduce safety
Amount of prestart safety training	Incomplete safety procedures
Providing a different type of safety training	Employee behavior which reduced safety
Relevance of safety training	Negative attitudes which reduce safety
Employees' failure to use safety training	Employees not following safety rules
Supervisors not supporting the use of safety training	Employees working under the influence of prohibited substances
Excessive (unsafe) noise in the workplace	
Excessive (unsafe) dust or fumes in the workplace	
Inadequate (unsafe) lighting in the workplace	
Precautions to prevent hazards occurring	
Faulty or unsafe equipment	
Out of date or old equipment	
Equipment maintenance	
Equipment which was unsafe to use	
Lack of equipment to do the job safely	
Being asked to operate equipment without sufficient training	
Lack of safety equipment	
Employees not using safety equipment	
Poor quality safety equipment	
Failures to enforce the use of safety equipment	
Inadequate safety inspections	
Outside contractors creating hazards	
Clients/customers creating hazards	
Work speed pressure from co-workers which reduced safety	

highlight issues which management did not know about. The information collected can be used to take corrective action and/or inform new employees of the risks.

9.5 Employee Perceptions and Exposure to New Employee Risk

Throughout this book, it has noted that new employees not only manage to kill themselves, but also pose a serious safety threat to other employees. Employees need to be very clear about the safety risks associated with new employees, and an organization should have a strategy in place which ensures all employees are provided with information on new employee risks. Part of this strategy, which may be included in a safety training program, should be to examine employees' attitudes toward organizational processes associated with the acquisition and onboarding of new employees, and how these attitudes expose the employee to risk. In order to achieve this, the organization would need to have employees complete measures of trust in selection processes (see Sect. 9.5.1), trust in prestart training process (see Sect. 9.5.1), new employee safety risk (Burt et al. 2009 used a single item to assessing the perceived risk associated with a new employee: *The risk of an accident/incident increases when a new employee joins my crew*), and compensatory behaviors (see Sect. 9.5.1). Employees' responses to these measures could be correlated, and the results presented to employees along with the results from similar analysis (e.g., the results shown in Tables 5.1 and 6.1 in Chaps. 5 and 6, respectively).

Table 9.10 Considerate and responsible employee (CARE) scale items	Scale items
	Workers should point out hazards to co-workers
	Workers should immediately remove hazards if possible
	Safety depends on everyone following safety procedures
	Co-workers should be warned when their actions are unsafe
	Workers should assist each other with tasks to ensure safety
	Co-workers should discuss changes that could improve safety
	Crew leaders should be notified of hazards
	Safety comes from worker cooperation
	Co-workers' limitations should be recognised
	Co-workers should give each other informal safety instruction
	Supporting co-workers ensures everyone's safety
	A worker should never be too busy to help a co-worker
	Co-workers should discuss near-hits
	Co-workers should discuss past accidents
	Near hits should be reported to management

9.5.1 Employee Perceptions of Organizational Processes: Selection, and Training

Burt and Stevenson (2009) and Burt et al. (2009) examined employees' perceptions of organizational processes and how these are associated with their reactions to new employees. Perceptions of recruit and selection processes are discussed in Chap. 5, and perceptions of socialization and prestart training processes are discussed in Chap. 6. Both chapters note how a perception that organizational processes helps ensure safety can be associated with a lowering of risk perceptions, a decrease in behaviors which should ensure safety, and thus an overall increase in workplace safety risk. Scales used to measure perceptions of trust in selection process, trust in induction processes, and employees' reactions to new employees (compensatory behaviors) are published in the appendix of Burt et al. (2009).

Burt and Stevenson (2009) reported Cronbach's alphas of 0.67 for trust in selection and 0.67 for trust in induction, while Burt et al. (2009) reported Cronbach's alphas of 0.76 for trust in selection, 0.72 for trust in induction, and 0.70 for compensatory behaviors.

9.5.2 Measuring Co-worker Caring

Given there will often be limitations associated with a supervisors ability to help a new employee adaption (as discussed in Chap. 4, Sect. 4.2.8), it is likely that an organization will have to formally assign a co-worker(s) to take on some of the responsibility for helping a new employee adapt and familiarize. While this responsibility should be officially assigned to a co-worker, a workers general attitude toward their co-workers safety will to some degree influence their commitment to, and engagement with, new employee adaption. When workers really do not care about their co-workers safety, safety risks for new employees are likely to increase. Thus, understanding the attitude of workers toward new employee (co-worker) safety is an important step in ensuring new employee safety. When it is evident that a work environment is characterized by little or no consideration and responsibility for coworker safety at the worker level, new employee adaption should not solely (formally or informally) be put into the hands of co-workers.

Burt et al. (1998) developed the *considerate and responsible co-worker CARE scale*. The CARE scale measures an employee's attitudes toward safety and particularly their attitude toward the safety of their co-workers. As such, the items tend to use the word *should*. An employee who holds a positive and caring attitude toward safety should tend to agree with the items presented in the scale. A model examining a number of variables which are associated with CARE development is presented in Burt et al. (2008). Burt et al. (1998) reported a coefficients alpha of 0.91, and a test–retest reliability of 0.62 for the original 21 item Care scale (also see Burt 2001). Burt, Chmiel, and Hayes found an acceptable co-efficient alpha

(alpha = 0.92) for a 15 item version of the CARE scale. Table 9.10 shows the 15 item version of the CARE scale. A worker who scores high on the CARE scale might be a good person to assign to a new employee to help with their adaption (assuming of course they also have the required knowledge).

9.5.3 Measuring Perceptions of New Employee Safety Behavior

In a work situation where there is a relatively high rate of new employee arrival, employees will develop perceptions of how new employees behave in relation to safety. From a very broad perspective, this could range from new employees who are typically unsafe to the point of being reckless, through to new employees typically behave very safely. Where on this continuum that an employee's perception of the safety behavior of new employees falls will probably determine how they are likely to act around new employees. If, for example, they feel that new employees typically behave very safely, they are perhaps less likely to monitor for dangers from new employees. In contrast, if the perception is that new employees can be reckless in regard to safety, employees' inclination should be to be very careful of the danger they may pose.

Clearly, an organization does not want to employ individuals that will be a danger to others. However, an organization also needs it workers to appreciate the dangers that can be associated with new employees. Furthermore, it would be useful for employees to understand how their perceptions of the typical new employee may be putting them in danger. The ideal situation is to have new employees that always behave safely, yet a workforce that is very weary of new employee safety

Table 9.11 Items to measure the perceived safety behavior of new employees

Scale items—retrospective	Scale items—prospective
Workers that have joined my team have had a positive attitude toward safety	Workers that join my team will have a positive attitude toward safety
Workers that have joined my team have worked in a safe way	Workers that join my team will work in a safe way
Workers that have joined my team have been open to talking about safety	Workers that join my team will be open to talking about safety
Workers that have joined my team have worked carefully as they were learning the job	Workers that join my team will work carefully as they are learning the job
Workers that have joined my team have been open to constructive criticism about their safety behavior	Workers that join my team will be open to constructive criticism about their safety behavior
Workers that have joined my team have readily asked for clarification on safety matters	Workers that join my team will readily asked for clarification on safety matters

behavior. Thus, workers that complete the prospective scale shown in Table 9.11 should tend to disagree with the items.

Burt and Hislop (2013) developed the 6 items shown in the left hand column of Table 9.11 to measure employees' perception of the safety behavior of past new recruits and reported a Cronbach's alpha for the 6 items of 0.79. The items in the right hand column of Table 9.11 are the prospective equivalent. These items measure worker expectations of new employees. Generally, expectations are based on past experience; therefore, an employee's responses to the two scales shown in Table 9.11 will probably be highly correlated. Of course, there are always exceptions, and while all previously encountered new employees may have been perfect safety citizens, the next new employee may be a real danger. Using the scales shown in Table 9.11, employees can be instructed (shown) how their expectations are driven by their previous experience, and also instructed in the dangers of making the assumption that all new employees are alike.

9.6 Conclusions

New employees will come to a workplace with a set of expectations, and some of these will relate to safety. Employees in the workplace will also hold attitudes and expectations about new employees and about organizational processes associated with the arrival of new employees. A new employee induction program can explain to new employees the nature of their expectations and how these are perhaps distorted, and can place them at risk. However, a far better way to deal with the safety risks associated with new employee expectations is to measure expectations and also acquire information from current incumbents and feed the information back to new employees. This way they can understand exactly how their views and their expectations compared to what job incumbents are saying. New employees also need to be informed about all the safety issues associated with the job they are entering and informed about the job's safety risk profile. While much of this information should be readily available, using a safety specific exit survey process may capture additional information which can either be used to correct the issue or used to alter the incoming employee to the issue.

It is also important to ensure that employees in an organization understand the risks that can be associated with new employees. Generally, people are at least a little weary of strangers, but it is clear that workers do place a lot of trust in organizational processes to protect them from new employee associated risks. The assumption that organizational processes will always lower risk associated with new employees is of course rather dangerous. The scales in latter part of this chapter offer measures which job incumbents can complete (perhaps during annual safety training), and which provide for information to be communicated back which should help increase employees' appreciation of the safety risks associated with new employees.

References

Barling, J., Loughlin, C., & Kelloway, E. K. (2002). Development and test of a model linking safety-specific transformational leadership and occupational safety. *Journal of Applied Psychology, 87*(3), 488–496.

Bjerkan, A. M. (2010). Health, environment, safety culture and climate-analysing the relationships to occupational accidents. *Journal of Risk Research, 13*(4), 445–477.

Burt, C. D. B., Banks, M., & Williams, S. (2014). The safety risks associated with helping others. *Safety Science, 62*, 136–144.

Burt, C. D. B., Gladstone, K. L., & Grieve, K. R. (1998). Development of the considerate and responsible employee (CARE) scale. *Work and Stress, 12*(4), 362–369.

Burt, C. D. B., & Hislop, H. (2013). Developing safety specific trust in new recruits: The dilemma and a possible solution. *Journal of Health, Safety and Environment, 29*(3), 161–173.

Burt, C. D. B., & Stevenson, R. J. (2009). The relationship between recruitment processes, familiarity, trust, perceived risk and safety. *Journal of Safety Research, 40*, 365–369.

Burt, C. D. B., Chmiel, N., & Hayes, P. (2009). Implications of turnover and trust for safety attitudes and behaviour in work teams. *Safety Science, 47*, 1002–1006.

Burt, C. D. B. (2001). Considerate and responsible employee (CARE) scale. In J. Maltby, C. A. Lewis, & A. Hill (Eds.), *Commissioned reviews of 250 psychological tests* (Vol. 2, pp. 804–805). UK: Edwin Mellen Press.

Burt, C. D. B., Cottle, C. K., Naswall, K., & Williams, S. (2013). Capturing safety knowledge: Using a safety-specific exit survey. In *Proceedings of the 14th European Conference on Knowledge Management* (Vol. 1, pp 99–105). Kaunas: Lithuania.

Burt, C. D. B., Sepie, B., & McFadden, G. (2008). The development of a considerate and responsible safety attitude in work teams. *Safety Science, 46*, 79–91.

Burt, C. D. B., Williams, S., & Wallis, D. (2012). New recruit safety expectations: Relationships with trust and perceived job risk. *Safety Science, 50*, 1079–1084.

Chmiel, N. (2005). Promoting health work: Self-reported minor injuries, work characteristics, and safety behaviour. In C. Krunka, P. Hoffmann, & A. Bussing (Eds.), *Change and quality in human service work* (pp. 277–288). Mering: Rainer Hampp Verlag.

Christian, M. S., Bradley, J. C., Wallace, J. C., & Burke, M. J. (2009). Workplace safety: A meta-analysis of the roles of person and situation factors. *Journal of Applied Psychology, 94*(5), 1103–1127.

Clarke, S. (2006). The relationship between safety climate and safety performance: A meta-analytic review. *Journal of Occupational Health Psychology, 11*(4), 315–327.

Costa, A. C., & Anderson, N. (2011). Measuring trust in teams: Development and validation of a multifaceted measure of formative and reflective indicators of team trust. *European Journal of Work and Organizational Psychology, 20*(1), 119–154.

Cox, S., & Cox, T. (1991). The structure of employee attitudes to safety: A European example. *Work and Stress, 5*(2), 93–106.

Diaz-cabera, D., Hernandez-Fernaud, E., & Isla-Diaz, R. (2007). An evaluation of a new instrument to measure organizational safety culture values and practises. *Accident Analysis and Prevention, 39*, 1202–1211.

Hayes, B. E., Perander, J., Smecko, T., & Trask, J. (1998). Measuring perceptions of workplace safety: Development and validation of the work safety scale. *Journal of Safety Research, 29*, 145–161.

Mueller, L., DaSilva, N., Townsend, J., & Tetrick L. (1999). *An empirical evaluation of competing safety climate measurement models*. Paper presented at the Annual Meeting of the Society for Industrial and Organizational Psychology, Atlanta, GA.

Neal, A., & Griffin, M. A. (2006). A study of the lagged relationship among safety climate, safety motivation, safety behavior, and accidents at the individual and group levels. *Journal of Applied Psychology, 91*(4), 946–953.

Neal, A., Griffin, M. A., & Hart, P. M. (2000). The impact of organizational climate on safety climate and individual behavior. *Safety Science, 34*(1–3), 99–109.

Sneddon, A., Mearns, K., & Flin, R. (2013). Stress, fatigue, situation awareness and safety in offshore drilling crews. *Safety Science, 56*, 80–88.

Tucker, S., Chmiel, N., Turner, N., Hershcovis, S., & Stride, C. B. (2008). Perceived organizational support for safety and employee safety voice: The mediating role of co-worker support for safety. *Journal of Occupational Health Psychology, 13*(4), 319–330.

Walker, A., & Hutton, D. M. (2006). The application of the psychological contract to workplace safety. *Journal of Safety Research, 37*, 433–441.

Zohar, D. (2000). A group-level model of safety climate: testing the effect of group climate on microaccidents in manufacturing jobs. *Journal of Applied Psychology, 85*(4), 587–596.

Chapter 10
Integration of the New Employee Safety Risk Management Processes

10.1 Introduction

The aim of this chapter is to provide an overall guide as to how the various management steps, recommendations, and processes discussed throughout the book can be adopted to enhance new employee safety, and indeed the safety of all employees. Organizations vary greatly in terms of the complexity of their human resource activities (e.g., complexity of documentation associated with a job, complexity of the processes used to recruit and select new employees, extent of induction and socialization processes, nature of prestart training, and nature of ongoing safety training), and their commitment and ability to adopt and deliver a high-performance safety management system. This chapter will attempt to explain how variation in the complexity of an organization's human resource activities may require adjustments to the recommendations made to manage new employee safety risk factors and how new employee safety can be managed within a relatively limited human resource management structure. While this chapter provides an overview of the new employee safety management recommendations made throughout this book, it is recommended that the full description of each safety issue and it associated management recommendations are examined before any organizational intervention is undertaken.

10.2 Step 1: Acknowledgement and Responsibility

There is little doubt that a new employee, particularly during their initial period of employment (roughly the first 3 months of employment in a new job), is at an increased risk of having an accident, and their presence in the workplace increases the safety risk for other members of an organization (see Chap. 2 for a review of the research evidence). Thus, the first step in managing new employee safety is for the

© Springer International Publishing Switzerland 2015
C.D.B. Burt, *New Employee Safety*,
DOI 10.1007/978-3-319-18684-9_10

organization to acknowledge that every new employee will bring an array of safety risks into the workplace. This acknowledgment should be formal, and the organization's safety officer(s) (or manager) should be specifically tasked with addressing all the safety issues associated with new employees. Thus, it is essential that the responsibility for new employee safety management is formally assigned to an individual or individuals.

While most of the safety issues associated with new employees are universal (always present), some will vary in terms of relevance as the nature of the new employee's job and workplace varies. In particular, the degree of contact or interaction that a new employee has with other employees will change a number of risk factors. The individual responsible for new employee safety will be able to isolate the specific safety risks associated with their organization's new employees, and develop a tailored management strategy. The overall management strategy will have more success if it is tailored to the specific situation. Each step in the overall management of new employee safety will need to be integrated with other human resource activities, and this should also be part of the individual's responsibility. To be specific, the individual(s) responsible for managing new employee safety will need to advise employees who may be engaged in general human resource processes on how to alter what they are doing in order to help ensure new employee safety. Furthermore, the individual will need to have the authority to work with supervisors and employees to ensure that they also understand how their attitudes and behavior can influence new employee safety, how they can be exposed to safety risks from new employees, and how they can help to manage new employee safety.

10.3 Step 2: Understanding Job Risks Using Safety Risk Profiling

The safety risks associated with a job can be a very complex mix of system, environment, and human factors. Furthermore, these risks can change as systems and equipment age, as work is completed in different geographical locations, and as employees with different lengths of job tenure populate the workforce. It is vital that new employees have a clear and realistic understanding of all the safety risks associated with the job they are about to enter. Thus, information on job's safety risks should be routinely collected, and communicated to job applicants, and new employees before they start work. Communication processes are discussed in the next section. To be successful, communication must be delivering the correct information which focuses on all types of safety issues: system, environment, behavior, attitudes, workload, etc. While there are very sophisticated techniques for analyzing task risks, a broad analysis of a job's safety risks (broad in the sense that it covers all areas where risk may come from) is perhaps best provided by asking employees that are performing the job which the new employee is being recruited for.

Often the employment of a new employee is preceded by the resignation of an employee, and this employee should be required to complete a *safety-specific exit survey* as part of the exit process. Chapter 4 discusses the use of a safety-specific exit survey, and Table 9.9 in Chap. 9 provides examples of items which can be included in a safety-specific exit survey. These items cover not only risks from systems, the environment, and equipment, but also from employees' behavior and attitudes, and from management factors. Arguably, the exiting job incumbent is the best person to inform the organization about the job's safety risks. They should know all the safety aspects associated with the job, including aspects, such as how co-workers and supervisors are behaving, which are difficult, if not impossible, to identify via any other means. The exiting employee may also have very specific safety information about system issues which would be lost if not requested before they exit the job. Consider for example the job of bulldozer operator. A resigning bulldozer operator may well be able to inform the organization about a safety issue with the equipment they have been using (driving). This information may lead to corrective action, or may lead to the new operator (the new employee recruited into the bulldozer operator position) being informed of the issue during their induction or prestart training process. As another example, consider the situation where equipment is being operated in an environment, where after rain the terrain becomes particularly (and unusually) slippery. An exiting operator will know this, whereas a new employee, unless they are informed before they start, may not be expecting this added hazard.

The overall objective of Step 2 is to collect as much safety information about the job the new employee will enter as possible, which can be integrated into the process of recruiting a new employee, starting with the provision of information to job applicants during the recruitment phase, and ending with a full briefing on the job's specific safety issues at induction and prestart training. As will be discussed further below, complete and realistic hazard and risk information is vital for all new employees, even those with many years of previous experience. As discussed in Chap. 3, previous job experience does not totally protect a new employee. For example, a new employee may have driven heavy trucks for many years, hauling a range of materials; however, they will have never driven the specific truck which they will be asked to drive in the new job, nor perhaps will they have driven on the specific terrain associated with the job. Thus, there is value in providing job safety and hazard information to all new employees, and this value is enhanced when the information's degree of specificity is high.

10.4   Step 3: Communicating Safety Information to Job Applicants and New Employees

Safety and hazard information about a job is really only useful if it is communicated to the people that need to know it. For example, if an exiting employee indicates that care needs to be taken with a particular operational aspect of a piece of

Table 10.1 A multilevel model of safety information communication

Target	Information resolution	Method of delivery	Objective
Labor pool	General safety information	Job advertisement	Realistic safety preview: self-selection in or out of application process
Job applicants	Detailed safety information	Job description and person specification	Safety expectation setting
New employee	Job specific safety information	Induction and prestart training	Awareness of, and coping with, specific risks and hazards

equipment, or in a particular operating environment, the people that need to know this information are those directly involved with the equipment's operation and maintenance. As such, the information in the latter example really *does not* need to be delivered to job applicants (but it is relevant to the person that gets the job). It is, however, vital that information about safety is provided to job applicants during the recruitment and selection phase of hiring a new employee. To keep the process efficient, and yet deliver the desired safety outcomes, an organization should have a multilevel strategy for communicating safety information, starting with communication to job applicants and followed up by further communication to the individual employed (to the new employee).

Table 10.1 illustrates a multilevel communication approach. The target of the information is shown in the left-hand column, the level of resolution and method of delivery are shown in the middle columns, and the objective at each level is shown in the right-hand column. The level of resolution of the safety information refers to how specific it is. Job applicants need general (yet job relevant) safety information, whereas the person that becomes the new employee needs all available safety information, including very specific information. Indicating that the job has safety risks does not mean that the organization is not doing everything possible to control the risks. In fact, alerting individuals to safety risks and hazards is in itself a useful step in the processes of controlling risk and hazard exposure.

One of the key objectives associated with providing risk and hazard information is to help set realistic safety expectations. That is, the provision of safety and hazard information can form a *realistic safety preview*. Chapter 3 describes four different types of job applicant and described how the four types vary in terms of safety expectations and previous job experience. Job applicants classified as *school leaver* or *career transition* applicants will have the least amount of relevant job experience and are likely to have the most unrealistic safety expectations. Clearly, *safety expectation setting procedures* will be of most benefit to those job applicants. However, it would be unwise to forgo the use of a realistic safety preview, based on the assumption that the job applicant, because of their previous experience, would not benefit from the process. Tables 3.1 and 3.2, in Chap. 3, show that there are safety risks associated with all job applicants, including career-focused applicants, and these can be reduced by providing a realistic safety preview. Thus, it is suggested that an organization always include a realistic safety preview process when recruiting employees. The following three sections outline the general nature of a *realistic safety preview* and a *safety expectation setting process.*

10.4.1 Realistic Safety Preview: Job Advertisement

It is vital that new employees have realistic safety expectations, and the process of ensuring that new employees have realistic safety expectations should begin with the provision of information when the job vacancy is advertised. At the recruitment stage, starting with information in the job vacancy advertisement, it may be suffi- cient to note the safety risks and hazards associated with the job in general terms. Furthermore, a new employee will require certain knowledge, skills, and abilities (competencies) in order to perform a job safely, and these should also be noted in the recruitment material. Providing information on safety, or safety relevant com- petencies, in recruitment material should allow some individuals in the labor pool to decide that they do not wish to purse the job. Those that decide to make an application for the vacant job, and become a job applicant, should now have a clearer understanding that there are safety issues associated with the job. They will also begin to see that the organization is committed to safety.

There is considerable variation in how an applicant pool can be generated. For example, a job vacancy can be placed on an online job site, in a newspaper, or posted on a notice board. Irrespective of how a job is advertised, the common element is that a written description of more or less detail is generally produced. A job vacancy advertisement should cover the job title, what the key tasks and roles are, what competencies are required to perform the job, how further information (e.g., job description and person specification) can be acquired, and how to apply for the position. To include a realistic safety preview, safety factors should be noted in the job advertisement. To omit any mention of safety is to miss an opportunity to ensure new employee safety. Furthermore, it is simply wrong for an organization to assume that individuals in the labor pool will already know about a job's safety risks. It is not necessary to provide extensive and detailed statements about safety in a job advertisement. Rather statements, such as *There are safety risks associated with undertaking aspects of this job. This job is performed in a hazardous envi- ronment. Safety aspects associated with this job require employees to work with extreme causation. Safety compliance is a key performance indicator for this job. In order to maintain safety the successful candidate will have the following compe- tencies …*, are examples of what could be included in order to bring the safety factors associated with the job to the attention of interested individuals.

Communicating safety information in recruitment material is essentially pro- viding the first step in the realistic safety preview process. As with a realistic job preview, some individuals will select themselves out of the process based on the information provided during recruitment. Other individuals will begin to develop clarity around the actual safety risks associated with the job, and this clarity will help to ensure that their safety expectations are more realistic. It is vital that organizations understand that new employees are likely to adjust their behavior to the level of risk which they assume exists. If the new employee has unrealistic expectations about safety, about risks, about hazards, or about the ability of systems, procedures, and people to protect them, they are likely to expose themselves to the possibility of an accident.

10.4.2 *Realistic Safety Preview: Job Description and Person Specification*

As noted, a realistic safety preview included in a job advertisement may result in some individuals deciding not to apply for a job. Those individuals that continue with the application process (that become a job applicant) should be sent the job description and person specification documents associated with the job. Organizations will vary in terms of the extent of these documents. Clearly, safety is enhanced, and also job performance, if an individual knows what a job requires, and a job description is one document that helps provide this information. Furthermore, both safety and performance are enhanced if an individual has the knowledge, skills, and abilities required to perform the tasks, and the person specification document describes the knowledge, skills, and abilities (competencies) which are required to perform a job. As noted in Chap. 3, a job description document can include a safety preview section. Figure 3.1, in Chap. 3, provides an example of a realistic safety preview section which can be inserted into a job description document. If an organization does not have job description and person specification documents, they should consider developing them. If this is not possible, they should at least provide a safety description, such as that shown in Fig. 3.1, to job applicants. Again the objective is to ensure that a new employee has a realistic understanding of the safety factors associated with the job they are entering. The process of providing a realistic safety preview and setting realistic safety expectations should continue through the selection process via questions on safety expectations (see Sect. 10.5), and is completed with the selected new employee during their induction and prestart training (see Sect. 10.6).

10.5 Step 4: Selecting New Employees

The majority of the work associated with acquiring a new employee is associated with examining information provided by job applicants and a consideration of whether this information indicates that the applicant could successfully perform the job. The steps taken by an organization to obtain information to allow for the select of a suitable new employee will vary considerably. In some instances, it might simply involve a rather unstructured conversation between a job applicant and an employer. In contrast, more sophisticated selection systems will involve multiple stages (e.g., application blank completion or CV provision, structured interviewing, psychometric testing, etc.), with a systematic consideration of applicants' scores on a range of predictors assessed against predetermined cutoffs.

From a new employee safety perspective, there are two key considerations associated with selection processes. First, what abilities do the measures or predictors used have to provide information on the job applicants work related outcomes (e.g., the individual's ability to perform the job safely, their attitude toward safety, and/or risk taking). Second, what assumptions do employees in the workplace hold about the organization's selection processes, and how can these assumptions influence workplace safety. The next two sections examine these two issues in more detail.

10.5.1 Selection Measures

Chapter 5 provides an extensive discussion of a range of different predictors which can be used to select employees. As discussed is Chap. 5, there are very few selection measures which can be used to accurately and reliably predict an individual's safety behavior. Of course, there are a lot of ways to measure the knowledge, skills, and abilities which may be needed to help an employee work safely. Thus, an organization needs to give careful consideration to the development and/or acquisition of the measures it will use when selecting employees. In particular, organizations should empirically show (or be provided empirical evidence if purchasing a selection predictor) that each measure that is used is indeed capable of accurately predicting what it is designed to predict. For example, a test which is sold as a predictor of an employee's safety behavior should be shown empirically to actually predict safety behavior. That evidence is known as *criterion-related validity* and should be obtained for all selection predictors. While every effort should be directed toward ensuring that a new employee has the knowledge, skills, abilities, and attitude required to work safely, reality is likely to be very different, in that even a very sophisticated package of selection predictors may have serious limitations when it comes to the accuracy of the predictions it makes. Of course, as the quality (complexity) of the selection predictors reduces (moves closer to the unstructured conversational pre-employment chat), the *possibility* of predicting a new employees' safety behavior is removed.

While the use of inaccurate selection predictors can reduce the ability of an organization to ensure (predict) new employee safety, assumptions about what selection information means, particularly job applicant's previous work experience, can also be problematic for safety. Chapter 3 provides an extensive discussion of experience and its relationship with new employee safety. Experience is a complex construct that should not be measured by simply examining a job applicant's *cumulative job tenure* (how many months or years they have previously worked for). Table 3.3 in Chap. 3 provides examples of more specific questions which can be used to measure a job applicant's experience and discusses how these questions can be scored in order to determine the value of an applicant's experience for ensuring safety. Table 3.4 provides questions which can be used to assess and score a job applicant's safety expectations. Inclusion of these questions in the selection process extends the realistic safety preview process into the selection stage. Of course, the questions in Tables 3.3 and 3.4 can be used in an application blank, and/or in a structured employment interview. Furthermore, the acquisition of detailed information about a job applicant's previous experience allows for limitations associated with experience to be dealt with during a new employees' induction, prestart training, and with on-the-job supervision.

In summary, selection processes can help to ensure new employee safety if they clearly define the knowledge, skills, and abilities that are required to perform a job, and obtain or develop accurate predictors of these. Put simply if an organization selects an individual for a job that does not have the knowledge, skills, and abilities which are necessary to perform the job in a safe manner, there will be an increased chance that the individual (the new employee) will be involved in an accident. Of course, working safely is also partly dependent on the new employee's attitude toward safety and on their personality (see Chap. 5). Unfortunately, attitudes and personality are not easy to measure in an error-free way. In this regard, an organization should not assume that they have very much ability at all to predict safety-related attitudes or to determine much in the way of safety behavior based on personality profiling.

10.5.2 Assumptions About Selection Processes

The second factor associated with the selection of employees relates to the views which employees in an organization have about the processes which are used by their organization to recruit and select new employees. Employees will have varying degrees of information about how their employer recruits and selects new employees. For example, they may know that the company advertises job vacancies on a specific Web site or in a local newspaper, and they may know the information provided in job advertisements. Employees may know what the company requests from job applicants, perhaps a CV or perhaps the completion of an application blank. Employees may know what additional information is provided to prospective job applicants, perhaps a job description and person specification. Employees may also know what selection steps are taken to find a suitable new employee and know whether these steps are simply an unstructured chat or a more sophisticated application of a range of selection predictors. Employees gain this knowledge through their own experience and through talking to other employees.

What employees are unlikely to know is the ability of their organization's recruitment selection processes to deliver an accurate outcome. Employee knowledge of this aspect of selection is largely an assumption. Chapter 5 provides an extensive discussion of research which has shown that as employees' trust in their organization's selection processes to deliver a new employee that will work safely increases, their perception of the safety risks associated with new employees will decrease, and at the same time, employees' engagement with new employees to ensure everyone's safety is reduced. Table 5.1 shows results from 3 studies which all show this basic pattern of results. Of course, this set of assumptions and relationship is very dangerous. If an employee's trust in recruitment and selection processes is misplaced, they will be wrongly assuming there is a lowered safety risk from new employees, and based on that assumption, they will not attempt to guard against the risks posed by new employees.

Thus, the second strategy associated with selection which will help with new employee safety is to ensure that employees understand the limitations of their organizations' recruitment and selection processes, and most importantly, how their assumptions about the organization's recruitment selection processes can potentially put them at risk. Ideally, during annual safety training (see Sect. 10.8), employees will be reminded of the risks associated with the results shown in Table 5.1, and how these translate into a sequence of assumptions which end in a decrease in employees ensuring their safety when a new employee arrives in the workplace. Furthermore, achievement of a *state-of-the-art* recruitment and selection system, one which is accurate and reliable, does not remove the necessity to inform employees that it is dangerous for them to assume that the recruit and selection system will deliver new employees that can be trusted to work safely. As discussed in Chap. 7, all new employees will go through an adoption process in their initial period of employment, during which time they will be a safety risk, both to themselves and to other workers. Thus, the careful monitoring of all new employees, by all members of an organization, is essential during the new employee's initial period of employment.

10.6 Step 5: New Employees' Induction and Prestart Training

Once a job applicant has been offered a position, they become a new employee. However, before they actually begin work, they should complete induction and prestart training processes. As with recruitment and selection processes, there are two key factors associated with induction, socialization, and prestart training processes which influence new employee safety. First, there is the issue of what is covered during the processes. Second, there is the issue of how employees within the organization perceive the processes. Induction and prestart training are likely to (or should) deliver a significant amount of general and safety-specific information to a new employee. These processes should also provide realistic safety preview information (see Sect. 10.6.1), discuss the principles of *a safety conscious helping culture* and the components of *a think before you help process* (see Sect. 10.7.3), and introduce the trust building process outlined in Sect. 10.7.2. Chapter 6 discusses induction, socialization, and prestart training processes in detail.

10.6.1 Setting Realistic Safety Expectations

Steps described in Sects. 10.4.1 and 10.4.2 attempt to ensure that new employees have realistic safety expectations and have a detailed understanding of the safety risks and hazards associated with the job they are applying for. However, it would be unwise to solely rely on the provision of information during recruitment (e.g., in a job advertisement and in a job description document) and selection as the mechanisms to ensure a new employee has realistic safety expectations. In addition to these steps, the inclusion of a safety expectation setting procedure as part of the new employee's induction process should add significant safety advantage. A safety expectation setting procedure that can be used during new employee induction is described in Chap. 3, Sect. 3.7.2. This procedure uses several scales to measure a new employee safety expectation, and these scales are shown in Chap. 9. The procedure involves the measurement of a new employees' safety expectations and feedback of their responses (using a format such as that shown in Table 3.5, Chap. 3) along with the results from current job incumbents. This simple process will allow a new employee to clearly see how their safety expectations compare with current job incumbents.

10.6.2 Acquiring Information During Induction and Prestart Training

Clearly, the objective of induction and prestart training is for new employees to learn and retain information. Unfortunately, this objective may not be achieved. Chapter 6 provides an extensive discussion of the limitations associated with training to effectively deliver information to new employees, and also how induction and prestart training can be designed to help improve their ability to effectively deliver information. Adopting the recommendations provided in Chap. 6 should help improve induction and prestart training outcomes. It is also extremely important that every new employee is assessed to determine that they have actually learnt the required material, and they should be required to pass an *evaluation process* before they are permitted to begin work. The objective of the evaluation process is twofold. First, it is important that the new employee does in fact learn and retain the material which is delivered during induction and prestart training, as it should help ensure their safety. But equally important is the possibility that other members of an organization will assume that new employees have learnt information during induction and prestart training. If this assumption (discussed in the next section) is incorrect, it has many safety implications.

10.6.3 *Assumptions About Induction and Prestart Training Processes*

Traditionally, induction and prestart training processes will focus to varying degrees on ensuring that new employees understand organizational rules and policy, are introduced to work procedures, or perhaps shown how systems and equipment are operated. As noted, Chap. 6 discusses research which has examined the ability of prestart training to deliver positive outcomes, but the chapter also discusses how employee's perceptions of prestart training can be associated with their trust in new employees to work safely, and how this trust tends to be associated with a lowering of perceived risk from new employees and a lowering of engagement with new employees to ensure everyone's safety. Thus, in the same way that employee's assumptions about a selection system can have a negative impact on safety, employee's assumptions about the effectiveness of prestart training (if they are incorrect) can also have a negative impact on safety. Unfortunately, prestart training is rarely evaluated, and studies which have examined training evaluation results often indicate very poor outcomes. To overcome this problem, an organization can (1) design induction and prestart training to maximize its effectiveness, (2) evaluate the outcome for each new employee, (3) only allow a new employee to begin work once they have passed the evaluation process which establishes that they have learnt the required material, and (4) communicate information on training limitations to job incumbent's during annual safety training (see Sect. 10.8).

10.7 Step 6: The Initial Employment Period

After a new employee completes induction and prestart training they will enter their new job, and will enter the initial employment period. The initial employment period, roughly the individuals first 3 months in the job, is an extremely dangerous time. During this time, a new employee can be exposed to numerous safety risks and must undergo adaption, familiarization, and trust development processes which have associated risks. Chapters 4 and 7 provide extensive discussion of these risk factors. To help ensure new employee safety (and indeed the safety of all employees), an organization needs to understand each risk factor and specifically manage it. Table 10.2 provides a summary of the safety risks discussed in Chap. 4, briefly notes the recommended management strategy, and notes where in Chap. 4 and in Chap. 7 a detailed discussion of each issue is located.

There will be some variation associated with each safety issue's noted in Table 10.2 applicability to a job. For example, the first three rows of Table 10.2 deal with equipment issues, and not every job requires the use of equipment. Similarity issues such as task assignment and working hours may have little flexibility associated with them. However, managing each applicable safety issue to the

Table 10.2 Safety issues to manage in the initial employment period

Safety risk issue	Management strategy	Detailed discussion in section
Equipment's operational risk	Assign new employee to operate least risky equipment	4.2.2
Equipment age	Assign new employee to operate new equipment	4.2.2
Equipment risk-based maintenance analysis	Assign new employee to operate equipment with low failure probability	4.2.2
Task assignment	Assign new employee to perform low-risk tasks	4.2.3 and 7.3
Working hours	Restrict to 8-h shifts	4.2.4
Performance expectations and workload	Control performance expectations and workload to allow adaption and avoid safety violations	4.2.5
Operating environment	Assign new employee to work in lowest risk operating environment	4.2.6
Team and co-worker characteristics	Place new employee into an experienced team	4.2.7
Supervision	Supervisor guidance, training, and oversight for an adequate length of time	4.2.8 and 7.4
Safety voicing	Develop and reinforce a safety voicing culture	4.2.9

best of the organization's ability, within the constraints of their, and the jobs, operational limitations should dramatically improve new employee safety. It is also important to note that the strategies listed in Table 10.2 are particularly important during the initial period of employment and during the time that the new employee is adapting, gaining familiarity, and developing trust relationships (see Chap. 7).

10.7.1 Familiarization

Chapter 7 provides a detailed discussion of the adaption and familiarization processes which a new employee will undergo during their initial employment period. These processes will place demands on new employees beyond those associated with their job and can potentially reduce the new employee's ability to maintain situational awareness. Organizations will vary in terms of how they handle new employee adaption and familiarization. However, adaption and familiarization processes can be managed to reduce the time required for their completion and thus remove the burden of these additional demands as quickly and as safely as possible. Chapter 7 notes how supervisors should be formally tasked with helping new employees adapt and familiarize. Chapter 7 also notes issues which may limit supervisor's ability to undertake these tasks, and how co-workers will often be involved in the process of helping a new employee adapt and familiarize.

Additional safety issues arise when co-workers enter into the process of new employee adaption and familiarization. If involvement of co-workers in the process is not managed correctly, there can be a significant increase in safety risks associated with helping reciprocity (see Sect. 10.7.3). For safety to be maintained, the use of co-workers to facilitate new employee adaption and familiarization needs to be a formally established relationship between the new employee and a co-worker(s). Furthermore, issues such as co-worker distance, performance demands, and occupational/job characteristics (e.g., physical distances and protective equipment use) need to be taken into consideration. Chapter 7 discusses each of these issues in detail.

10.7.2 Trust Development

Trust plays a central role in workplace safety. For example, the organization trusts employees to perform tasks in a safe way. Employees trust management and co-workers to ensure their safety. Chapter 7 provides an extensive discussion of trust. Without doubt, an organization will function more effectively and safely if relationships between management and employees, and between employees, are characterized by trust. However, there are negative aspects to trust. Trust tends to reduce a person's monitoring of others behavior, and when it comes to safety, this can be very dangerous. Chapters 5 and 6 discuss in detail how employees can trust new employees to work safely based solely on their perception of their organizations' recruitment, selection, and training processes. Clearly, this is very dangerous. In my view, trust needs to be earned, and everyone's safety will be improved if an organization formally adopts a systematic data-driven process by which trust relationships between new employees and other members of the organization are developed. Section 7.7.1, in Chap. 7, describes such a process. The process is relatively simply to adopt, and a team can easily be trained in its use.

10.7.3 Helping Behaviors

Chapter 8 discusses helping behaviors and offers a number of examples of how helping can result in an accident. Unfortunately, there are many reasons why new employees may wish to engage in helping during their initial period of employment, and these are also outlined in Chap. 8. Three strategies are suggested to reduce the likelihood of new employees being involved in an accident, or causing an accident, because of attempts to help. The first is to use the *consequences of helping scale* (see Chap. 9, Sect. 9.3) during the new employee's induction process. Completion of the *consequences of helping* scale and a discussion of the new employee's responses should alert the new employee to the risks associated with helping. Chapter 8 described two other strategies: development of *a safety conscious helping culture* within the organization and training new (all) employees in *a think before you help process*. Section 8.7.1 offers 8 principles which could form the key components of a *safety conscious helping culture*. Ideally, new employees would be instructed in these during their induction process. In conjunction with this, the new employee's induction process should train them in the *think before you help process* described in Sect. 8.7.2. Overall, the objective is to ensure that new employees understand the safety risks associated with helping and carefully evaluate all aspects of a situation and the implications of their actions, before deciding to engage in helping.

10.8 Step 7: Annual Safety Training

All employees (including management) need to be reminded of the safety risks associated with new employees, and the nature of, and reasons for, the implementation of the management strategies used to ensure new employee safety. All

Table 10.3 New employee safety management strategies and issues to discuss during annual safety training

Management strategy	Knowledge of the risks and safety advantages
The reasons for the use of realistic safety preview processes during recruitment, selection, and induction	The safety risks if new employees have unrealistic safety expectations
The emphasis on a multilevel assessment of experience during recruitment and selection	The safety risks associated with assuming a new employee's previous experience makes them experienced
Why the organization uses a safety-specific exit survey process	The safety risks if a new employee does not know all the risks and hazards associated with their specific job
Organizations' commitment to a safety voicing culture	Why new employees may initially be reluctant to voice safety concerns, and the risks this may pose
New employee adaption and familiarization is a part of the supervisors' job, and there may be a co-worker(s) formally assigned to help	How informally helping a new employee can promote new employee *helping reciprocity* which has many associated risks
The principles of the *safety conscious helping culture,* and the reason for training in the *think before you help process*	Safety risks associated with helping, and reasons why employees in their initial period of employment are instructed not to engage in helping
Organization to communicate realistic information about the ability of recruitment, selection, prestaring training and induction processes to ensure new employees will work safely	Trusting organizational processes to deliver new employees that will work safety, based on assumptions about their effectiveness, is extremely dangerous
New employees initial period of employment must be characterized by a careful consideration and management of the factors shown in Table 10.2	Management of the factors listed in Table 10.2 will improve safety for all organizational members
The reasons for the operation of the trust development process	The risks associated with blindly given trust, and trust reduced monitoring of new employee behavior

employees should also be reminded how their behavior can influence new employee safety. An ideal place to discuss all of these issues in during annual safety training (in fact, safety training focusing on new employee risks should perhaps occur more frequently if a lot of new employees are arriving). However, employees should, at least once a year, be reminded of the key points shown in Table 10.3. The left-hand column of Table 10.3 notes the new employee safety management strategy (which would be explained to employees), and the right-hand column notes the key risk aspect(s) which it addresses (which should also be explained to employees).

10.9 Conclusions

This chapter has provided a brief description of the key points made throughout this book in a single source. Arguably, improving new employee safety requires the adoption of a set of relatively straightforward strategies. Of course, the adoption of the recommendations requires a commitment on the part of the organization to invest time and resources in new employee safety. It is also clear that co-workers have a large part to play in new employee safety. Finally, it seems clear that if new employee safety is not actively managed, new employees will continue to have accidents and be killed at disproportionally high rates.

Index

© Springer International Publishing Switzerland 2015
C.D.B. Burt, *New Employee Safety*,
DOI 10.1007/978-3-319-18684-9